GREAT INDIAN PHILOSOPHY

GREAT INDIAN PHILOSOPHY

Editor-in-Chief
Swami Chidatman Jee Maharaj

ANMOL PUBLICATIONS PVT. LTD.
NEW DELHI - 110 002 (INDIA)

ANMOL PUBLICATIONS PVT. LTD.
H.O.: 4374/4B, Ansari Road, Daryaganj,
New Delhi-110 002 (India)
Ph.: 23278000, 23261597
B.O.: No. 1015, Ist Main Road, BSK IIIrd Stage
IIIrd Phase, IIIrd Block,
Bangalore - 560 085 (India)
Visit us at: www.anmolpublications.com

Great Indian Philosophy

First Published, 2009
ISBN 978-81-261-3616-2

PRINTED IN INDIA

Printed at Mehra Offset Press, Delhi.

Contents

Preface

India colourful and vibrant, a land as diverse as its people. A mosaic of faiths, cultures, customs and languages that blend harmoniously to form a composite whole. One of the world's oldest living civilizations—which gave to the world—the concept of zero, the primordial sound Aum..., Yoga and Buddhism.

She is old. She is young. The two diametrically opposite, different yet unique and dynamic faces of India coexist even at this turn of the millennium. With a history of heritage, learning and civilisation dating back to almost 5000 years or perhaps even more, the modern India presents itself as an ideal country. From snow clad mountains to breezy sparkling blue sea side. From arid deserts with shifting sand dunes to lush green country sides. From ancient relics and architectural ruins dating back into antiquity to modern cities, India has it all.

To the north it is bordered by the world's highest mountain chain, where foothill valleys cover the northernmost of the country's maximum states. Further south, plateaus, tropical rain forests and sandy deserts are bordered by palm fringed beaches. Its southernmost tip is the meeting point of three Oceans.

India is a land of staggering contrast with a mingling of the tradition and modernity that is an unique experience to be in India. One will forever remember the stay in this wonderful Homeland-a land truly of mystery and enchantment. Only India can offer such an astonishing variety of contrasts.

She has given birth to numerous spiritual leaders and founders of some of the popular religions of the present day world. India probably has the most religious diversity in any country. It's the birthplace of Hinduism, Buddhism, Sikhism and Jainism and is among the few places to have a resident

Zoroastrian population. India with her veritable treasure trove of culture, mysticism, philosophy, art, music and architecture to name a few. Much of India's classical music is devotional and the North Indian Hindustani and South Indian Carnatic streams are distinct and both have a complex 'raga' framework. The legacy of dance in India is tremendous. The classical dances of India are numerous such as Kathakali of Kerala, Bharatnatyam of Tamil Nadu, Kuchipudi of Andhra Pradesh, Manipuri and Odissi from Orissa are the prominent dance forms in this country that sways to an altogether novel beat. The Yakshagana, nautanki and puppetry are ancient folk forms that live on till date. The earliest specimens of Indian painting are the ones on the walls of the Ajanta Caves dating back to 2nd century BC. The Mughals had a huge impact on Indian art. The influence of Persian art brought placid garden scenes, illustrations from myths, legends and history into Indian art. Word craft, handicrafts, architecture and sculpture all contribute to this rich and varied domain. Indian literature, both in English and in the vernacular, is ever more popular around the world.

The book aims at presenting perspectives of different aspects of Indian culture. It has topics on Indian classical music and dance, painting, ayurveda and yoga which are traditional and indigenous system of physical and mental wellbeing, language, philosophy and excellence achievements by Indian in the different branches of science and social sciences. It also includes our ancient religious and classic literature.

It is hoped that this book will be welcomed by the general readers as also scholars who are interested in the Indian heritage.

—Editors

1

Ancient Hindu Philosophy

Other Theories put Forward to Explain the Phenomenon: There are a number of small children who have reported having memories of past lives prior to their present life, and some also report being able to recall a time between lives (see books by Dr. Ian Stevenson, Carol Bowman, Dr. Jim B. Tucker, and Elisabeth Hallett). In some cases these children have also reported being in a place like heaven between lives, and sometimes that they were given some degree of choice as to whether and when to be reborn, and even in selecting their future parents.

Some of these children claim that being reborn is not necessarily a punishment for past bad "karma", but rather an opportunity for a soul to grow spiritually. Additional lifetimes could give individual souls a greater opportunity to accomplish more for God, if that is a person's goal, and to develop better character traits. Eastern views of reincarnation vary and several parallels with this idea are to be found in certain branches of Hinduism and Buddhism.

Samsara

In Hinduism, Buddhism, Jainism and other related religions, samsara or sacsara refers to reincarnation or rebirth in Indian philosophical traditions.

Etymology: Samsara is derived from "to flow together," to go or pass through states, to wander. Mostly a great revolving door between life and death and a new life reincarnated cycle of life. Also known as a game in ancient India.

Cycle of Rebirth: In most Indian philosophical traditions, including the Hindu, Buddhist and Jain systems, an ongoing cycle of birth, death, and rebirth is assumed as a fact of nature. These systems differ widely, however, in the terminology with which they describe the process and in the metaphysics they use in interpreting it. Most of these traditions, in their evolved forms, regard Sacsara negatively, as a fallen condition which is to be escaped. Some, such as Advaita Vedanta regard the world and Sacsaric participation in it as fundamentally illusory.

Some later adaptations of these traditions identify Sacsara as a mere metaphor.

Sacsara in Hinduism: In some types of Hinduism, Sacsara is seen as ignorance of the True Self, Brahman, and thus the soul is led to believe in the reality of the temporal, phenomenal world. In Hinduism, it is avidya, or ignorance, of one's true self, that leads to ego-consciousness of the body and the phenomenal world. This grounds one in desire and the perpetual chain of karma and reincarnation. The state of illusion is known as Maya.

Hinduism had many terms for the state of liberation like moksha, mukti, nirvana, and mahasamadhi.

The Hindu Yoga traditions hold various beliefs. Moksha may be achieved by love of Ishwar/God (see bhakti movement), by psycho-physical meditation (Raja Yoga), by discrimination of what is real and unreal through intense contemplation (Jnana Yoga) and through Karma Yoga, the path of selfless action that subverts the ego and enforces understanding of the unity of all. Advaita Vedanta, which heavily influenced Hindu Yoga, believes that Brahman, the ultimate Truth-Consciousness-Bliss, is the infinite, impersonal reality (as contrasted to the Buddhist concept of shunyata) and that through realization of it, all temporal states like deities, the cosmos and samsara itself are revealed to be nothing but manifestations of Brahman.

Sacsara in Jainism: In Jainism, karma, anuva (ego) and the veil of maya are central.

In Jainism, liberation from samsara is called *moksha* or *mukti*.

Samsara in Surat Shabda Yoga: In Surat Shabda Yoga, the purpose is to realize the individual's *True Self* (Self-Realization), *True Essence* (Spirit-Realization) and *True Divinity* (God-Realization) while living in the human physical body. This *Journey of Soul* involves reuniting in stages with what is called the *Essence of the Absolute Supreme Being*, the Shabd. Attaining self-realization and above also results in *jivan moksha / mukti*, liberation/release from *samsara*, the cycle of karma and reincarnation while in the physical body.

Surat Shabda Yoga Cosmology presents the constitution of the initiate (the microcosm) as an exact replica of the macrocosm. Consequently, the microcosm consists of a number of bodies, each one suited to interact with its corresponding plane or region in the macrocosm. These bodies developed over the yugas through involution (emanating from higher planes to lower planes) and evolution (returning from lower planes to higher planes), including by karma and reincarnation in various states of consciousness.

Tattva

According to the Indian school of Samkhya philosophy, the *Tattva* are a way of directly "experiencing" the 5 alchemical elements. So they are therefore the logical progression of the previous elemental grade where the elements were first discussed and symbolic ways of working with them described. The student is then taught a method of directly learning the "theoretical" structure of the elemental worlds.

A Brief History of the Tattvas: There are six main schools of thought on Tattvic philosophy. The original *Tattva* system was developed by the Indian Maharshi Kapila as part of his Samkhya philosophy. This was about 700BCE, but the roots of Tattvic philosophy go back to at least 2000BCE. Kapila's Samkhya philosophy system divides the universe into five basic Tattvic shapes, which when countercharged become 25 Tattvas.

The word Tattva is comprised of two words, *Tat* (meaning *that*) and *Tvam* (meaning *thou*). Tattva basicially means *thatness*, that is, the real being of anything. It is generally translated as meaning *quality*. Further, *Tat* represents the Godhead and *Tvam* the individual, giving a meaning (which fits in well with Hermetic Philosophy) of "That (which is the Universe) art thou." This is similar to the Hermetic axiom of "As above, so below", and is directly related to the concepts of the Macrocosm (*Tat*, Godhead) and Microcosm (*Tvam*, individual).

The Hatha Yoga School of Tattvic Philosophy links the energy found in breathing (Prana) with the cycle of the five Tattvas. The eighth chapter of the *Shivagama* is *The Science of Breath and the Philosophy of the Tattvas*. It is written within this work that "The Universe came out of the Tattvas; it goes on by the instrumentality of the Tattvas; it disappears into the Tattvas; by the Tattvas is known the nature of the Universe."

The Tattvas are the five modifications of the Great Breath, Prana, which is described as the life principle of the Universe (macrocosm) and man (microcosm). Prana consists of an ocean of the five Tattvas. The Buddhist texts of Indian Tantra describe 7 energy centres (chakras) in connection with the five Tattvas. This finds its ultimate refinement in the Tibetan Buddhist fivefold chakra system, which is part of their fivefold division of esoteric symbolism for categorising the Universe. The Tattvas correspond to the five lower chakras in the Indian scheme of energy centres, and to all five psychic centres of the Tibetan chakra scheme.

H.P. Blavatsky's Theosophical Society, based in Adyar, India, derived the bulk of their knowledge on Tattvas from Rama Prasad, who taught the Tattvic philosophy of the Hatha Yoga School. The Theosophists combined these teachings with those of the Esoteric Tibetan Buddhist tradition. On top of this they conjectured two more Tattvas, to make a total of seven, which allowed them to directly match the Tattvas to there sevenfold system, in particular to the seven Chakras.

The *Tattvas* they added were *Adi*, which is egg shaped, and *Anupadaka*, which is a crescent moon containing a radiant white sun. The Hermetic Order of the Golden Dawn incorporated the Theosophical Tattva system, however discarding the sixth and seventh conjectured Tattvas, The Golden Dawn utilised Kapila's Sankhya Tattva system of 25 countercharges. This was the only Eastern concept that the Golden Dawn introduced into their system. It was perhaps influenced by the mention of Tattvas in Bulwer-Lytton's novel *Zanoni*, which was very likely the first mention of Tattvas in Western literature. In *Zanoni* the poisoning of the main character was neutralised by the magical visualization of the *Apas* Tattva.

The Golden Dawn system parallels the five Tattvas to the Western Doctrine of the 5 elements, the four primary elements and quintessence.

The Tattvic Philosophy in Western Tradition: The neutrality of this section is disputed. Please see the discussion on the talk page. The Tattvas are the astral form of the Elements, upon which the Physical elements are based. The process of how the Tattvic elements become manifest is a complex process, which will be described briefly below.

In the Tattvic model a solar wind constantly streams forth from the Sun. It is a partially electro-magnetic nitrous gas, which is very subtle and non-physical in nature (experiment show that the solar wind is mostly composed of high-energy electrons and protons). This is the Prana (*Life Matter*) of the Eastern Mysteries. The Earth's magnetic field captures this incorporeal nitre as the wind streams past our planet. This unmanifest substance circulates around the planet in a series of fivefold waves, each of which comprises five sub-waves. These waves are constantly rising and falling. It becomes more and more physical as it passes through our atmosphere. A salt is formed as this nitre passes through water vapour. This is "a most subtle virgin earth," the Prima Materia of the alchemists. Chemically if you could isolate pure NO_3 it would not form a salt on contact with water.

The Tattva flows start at sunrise and flow from Spirit, to Air, to Fire, to Water, to Earth. Each sub-Tattva takes four minutes and 48 seconds to transit a given spot, meaning it takes a total of 24 minutes for each main Tattva. Thus it takes a total of two hours for a primary flow of all five Tattvas to transit. The Earth's electromagnetic field and ley lines are vitalised by these currents. Current Scientific models suggests that earths magnetic field is best explained by Dynamo theory.

All five Tattvas are considered to be *aethers*. The atoms of the *aether* bring the atoms of the atmosphere into the necessary state of vibration for us to perceive it. Each of the Tattvic aethers is related to one of the five senses. Rama Prasad gives the example that Light is the manifestation of the Tejas Tattva (the luminiferous aether) within the atmosphere, that is, matter is brought into the correct visual vibration for us to perceive it. However, physicists and biologists have shown that vision works through a process of photons reflecting off objects and being detected by cells in the eye.

In *Nature's Finer Forces*, Tattvas are defined as the following: "a mode of motion; the central impulse, which keeps matter in a certain vibratory state; a distinct form of vibration. Also, "every form and every motion is a manifestation of these Tattvas singly or in conjunction."

The first statement reflects what hermetic science states of the nature of the alchemical processes, that is, "*to raise the vibrations of matter*" and thus bring it into the condition of harmony, Quintessence. Modern Physics concurs with this ancient knowledge that all matter is energy at various vibratory rates.

The second statement, that all matter and motion is a manifestation of the Tattvas is a direct parallel with the Western Doctrine of Elements, which also asserts that all matter is comprised of varying proportions of the subtle elements.

The other direct parallel between the Eastern and Western systems is that Tattvic philosophy asserts that *Akasha* is the most important Tattva. *Akasha* is what Western Tradition calls *Spirit*, or Quintessence, and like the four base elements all deriving from *Spirit*, "it is out of Akasha that every form comes, and it is in Akasha that every form lives."

The Western Tradition's Doctrine of Elements is exemplified by the Enochian system of magic. Compared with the Tattvas, the Enochian system has been more intellectualised. This seems only natural in that (in psychological terms) the Western tradition is one of "thinking" and the Eastern tradition is one of "feeling."

Modern Enochian magicians have proposed a direct correlation between the Elements and the Forces of modern science. This is perhaps very relevant to the Tattvas, in that as "aethers" they are subtle and intangible, something that gives form to matter. Akasha/Spirit is correlated to the hypothesised "Superforce" of the Grand Unified Field theory of Relativity. Air and Fire are in juxtaposition in Enochian, compared to the Vayu and Tejas of Tattvic Philosophy, but are equated to the weak and strong nuclear forces of physics. Water/Apas is linked to electromagnetism and Earth/Prithivi is linked to Gravity.

Symbols and Colours: Each of the Tattvas has a corresponding symbol and colour. These symbols differ from traditional Western symbols in shape and colour, and also vary slightly between the various Tattvic philosophies. Each of the Tattvas and their attributes will be described separately below.

Akasha The Akasha Tattva is associated with the spirit of Western Tradition and is the sonoriferous aether (sound). It is represented by some as a black or indigo Vesica Piscis, which is almond shaped. It represents the womb of the Universal Mother. In other versions, such as the Golden Dawn tradition, Akasha is symbolised by an oval or egg shape. Akasha is "every colour" or "clear." Black can also be seen as the absorption of all other colours, and its "flashing" (complementary) colour is luminous.

The quote given before, that Akasha is the most important Tatva, in that all forms come from and live in it, is of particular importance. The Akasha Tattva, when used as a "doorway" for scrying can allow one to scan for information within what Theosophists have termed the Akashic Record. The Akashic record contains, as the quote states, all forms. It could be compared with the "collective unconscious" of Jungian psychology. It contains all memories of human experience, but beyond it contains all future "history" in seed form.

Vayu or Pavan Vayu or Pavan is the tangiferous aether (touch). It is associated with the element of Air in Western Tradition. It is represented by a sky-blue circle.

Agni or Tejas Agni or Tejas is the luminiferous aether (sight/light/colour). It is associated with the element of Fire in Western Tradition. It is represented by a red equilateral triangle. The Fire of Tejas is considered to be Agni, which is the Vedic sacrificial fire and the interior fire of Kundalini Yoga.

Aap Aap is the gustiferous aether (taste). It is associated with the element of Water in Western Tradition. It is represented in the Tibetan version as a purple half circle. The Golden Dawn depicted Apas as a white or silver crescent, with the two points up. Prithvi or Bhumi Prithvi or Bhumi is the odoriferous aether (smell). It is associated with the element of Earth in Western Tradition. It is represented by a yellow square.

The sub-Tattvas are represented by placing a smaller Tattvic symbol within the larger main Tattvic symbol. This gives a total of 25 Tattvas: Spirit of Spirit, Air of Spirit, Fire of Spirit, Water of Spirit, Earth of Spirit; Air of Air, Fire of Air, Water of Air, Earth of Air, Spirit of Air; Fire of Fire, Water of Fire, Earth of Fire, Spirit of Fire, Air of Fire; Water of Water, Earth of Water, Spirit of Water, Air of Water, Fire of Water; & Earth of Earth, Spirit of Earth, Air of Earth, Fire of Earth, Water of Earth.

Clairvoyance and the Tattvas Within the Western Tradition, the Tattvas are primarily used to help develop the faculty of astral clairvoyance.

The symbol of the Tattvas is placed upon cards. Sub-Tattvas are placed within the larger Tattva image. The process of scrying is started by sitting comfortably and gazing fixedly at the Tattvic symbol until it is thoroughly assimilated. The edges of the Tattvic symbol will appear luminous after several minutes, as the flashing (or complementary) colour appears on the retina. This complementary colour is considered to be the astral counterpart of the actual colour. While meditating on the chosen Tattva, one should also "feel" the attributes of the associated element, letting the entire consciousness be filled with the appropriate sensations. For example with the Vayu Tattva, one could imagine a breeze blowing around and through them, concentrating on how this airy breeze feels, smells, tastes and sounds.

Once the Tattva has been completely assimilated the card can be set aside and the eyes closed. The Tattvas symbol should appear in its flashing colour. This symbol serves as a doorway through which one travels. When entering through the symbol one should feel themselves pass through a 'veil' or membrane. Once on the other side, one should take a moment to adjust to their new surroundings, taking note of what the new surroundings consist of.

A guide can be summoned to accompany you on your exploration of the surroundings, who can explain aspects of the environment to you, and take you to important entities who live within the realm (According to 'safety' and 'protocol' various esoteric groups insist on the following: when first met, a guide should be tested with elemental signs, and the highest magickal names associated with the element. If the guide responds correctly to the signs, then it can be followed in good faith.) It is important to show all due courtesy to a guide, and thank it for its time and co-operation and bid it farewell. It is also of benefit to ask its permission to summon it again when you next return to this Tattva.

This initial entry point is also where one departs from, stepping backwards through the portal. Once back on the other side, one should visualise the portal closing.

This process is intended to allow one to enter into the actual element (Tattva flow), initially via "active imagination" and eventually by spontaneous clairvoyance. By scrying the astral current we are attempting to see beyond the material manifestation of an element by transmuting the physical Element into its astral state. The intent is to learn about the vital principle, which the element represents. Within the astral this vital principle becomes perceptible to us as living beings of approximately human form. Everything within this astral environment is of a form and appearance that is highly symbolic of the meanings and intentions of what it represents. On the physical plane humans have obscured this process of dressing in a ways not symbolic of their true natures. Most animals however still retain a degree of symbolism, such as bright colours often representing the dangerous nature of the animal (especially in the case of tropical fish, tree frogs, and spiders).

On the astral, Elementals may masquerade for a while in deceptive form, buts its true form is always constant with its Element and character. (More from the 'traditions': To try and 'force' them into assuming their true form one can test them by using God Names, Grade/Elemental signs, Invoking/ Banishing forms of the Elemental pentagrams). Deep scanning of an entity's core ideals and feelings allows one to compare how it "feels" with how it "appears," and if the feelings don't match with its outward appearance then deception can be detected.

Trimurti

In Hinduism, the Trimurti (also called *the Hindu trinity*) is a concept that holds that God has three aspects, which are only different forms of one God. The three aspects of God, or "Parabrahman," or God's *personae* are as Brahma, Vishnu and Shiva. This Trimurti concept is a tenet most strongly held in Smartism, a denomination of Hinduism and rejected by other denominations of Hinduism, such as Vaishnavism. For example, the Dvaita school holds Vishnu alone to be the supreme God and Shiva as subordinate to Vishnu and interpret the Puranas differently. For example, Vijayindra Tirtha, a Dvaita scholar interprets the 18 puranas differently. He interprets that the Vaishnavite puranas as satvic and Shaivite puranas as Tamasic and and that only satvic puranas are considered to be authoritative.

- Brahma-the Source/Creator
- Vishnu-the Preserver/Indwelling-Life
- Shiva-the Transformer (Destroyer-Creator)

The Trimurti itself is conceived of as a deity and artistically represented as a three-faced human figure. Brahma is no longer as relevant as He once was in Hindu writing; some would say that focus on the aspect of the all-in-one Devi or of Shakti, the Divine Mother or God's Power personified, has replaced focus on the Source/Creator as Brahma. According to the Trimurti belief, these three personae of God are simply different aspects of the one and the same God. In this manner, such beliefs are similar to some interpretations of the Christian Trinity such as Sabellianism.

Swami Sivananda, in his book, *All about Hinduism*, noted that "Brahma represents the creative aspect; Vishnu, the preservative aspect; and Shiva, the destructive aspect of Paramatman. These functions are akin to wearing different garbs on different occasions. For example, when you do the function of a judge, you put on one kind of dress. At home you wear another kind of dress. When you do worship in the temple, you wear another kind of dress. You exhibit different kinds of temperament on different occasions. Even so, the Lord does the function of creation when He is associated with Rajas Guna, and He is called Brahma. He preserves the world when He is associated with Sattva Guna, and He is called Vishnu. He destroys the world when He is associated with Tamas Guna, and He is called Shiva or Rudra."

A widely accepted belief is that it represents earth, water, and fire. The earth, or Brahma, is seen as the originator of all life and hence is regarded as the Creator. Water is the sustainer of life and hence is the Preserver and is represented as Vishnu. Fire transforms or consumes life and hence is the Destroyer and is represented as Shiva.

The Trimurti can also represent three individual forms on the different planes of consciousness. On the spiritual plane, the spiritual element is represented by Brahma, the mental/psychic element by Vishnu, and the physiological element by Shiva. On the mental/psychic plane, Brahma is the intuitive and creative thought, Vishnu is intelligence, and Shiva stands for emotions and feelings. The Sky as Brahma, the Sun as Vishnu and the Moon as Shiva form the earthly/physiological plane. Another representation by Adi Shankara, Shiva represents the Nirguna Brahman, Vishnu represents the Saguna Brahman and Brahma the Cosmic Mind.

Various phases of an individual's life are said to be represented by the Trimurti. The first of these phases, that of celibacy and studentship, (Brahmacharya Ashram) – is represented by Brahma. During this phase, knowledge is the individual's constant companion. Knowledge, in this conception, is represented by Goddess Saraswati, who is said to be the consort of Brahma. The second phase of adulthood and household (Grihastha Ashram) is represented by Vishnu. During this phase, the individual fulfils all religious and family obligations by involving oneself in generating wealth, which one uses to sustain oneself and one's family. During this phase wealth is the individual's companion and is represented by Vishnu's consort, Goddess Lakshmi.

The third phase is that of old age (Vanaprastha Ashram) and is represented by Shiva. This phase marks the renunciation of one's worldly life for a life void of material pleasures, dedicated to the pursuit of true knowledge. In ancient days, this typically marked the time when a householder, along with his wife, left their worldly belongings to spend their life in a forest, just like Lord Shiva leads a homeless life with only the essential belongings. The final phase (Sanyasa Ashram) the individual seeks to merge oneself with the Supreme power (Ishwara). One completes the process of renunciation and he along with his wife lead a life completely untouched by any attachments. The only occupation becomes deep meditation, leading to the individual becoming Ishwara Himself, with his wife as Ishwara's consort becoming the Universal Mother. The three phases of life, represented by the Trinity thus culminates into One underlining the fact that the three are in reality one and the same Ishwara.

In philosophical terms, Brahma is said to be associated with Divinity's Creative Ground of Being, Vishnu is said to be associated with Divinity's Emanated Idea (Logos, Wisdom, or Word), and Shiva is said to be associated with Divinity's Transformative Energy (Flame, Breath, or Spirit).

Evolution of Theology on the Hindu Trinity

The definite settlement of the caste system and the Brahmanical supremacy must probably be assigned to somewhere about the close of the Brahmanas period. Division in their own ranks was hardly favorable to the aspirations of the priests at such a time; and the want of a distinct formula of belief adapted to the general drift of theological speculation, to which they could all rally, was probably felt the more acutely, the more determined a resistance the military class was likely to oppose to their claims. Side by side with the conception of the Brahman, the universal spiritual principle, with which speculative thought had already become deeply imbued, the notion of a Supreme Personal Being, the author of the material creation, had come to be considered by many as a necessary complement of the pantheistic doctrine.

But, owing perhaps to his polytheistic associations and the attributive nature of his name, the person of Prajapati seems to have been thought but insufficiently adapted to represent this abstract idea. The expedient resorted to for solving the difficulty was as ingenious as it was characteristic of the Brahmanical aspirations. In the same way as the abstract denomination of sacerdotalism, the neuter brahman, had come to express the divine essence, so the old designation of the individual priest, the masculine term brahma, was raised to denote the supreme personal deity which was to take the place and attributes of the Prajapati of the Brahmanas and Upanishads.

However the new dogma may have answered the purposes of speculative minds, it was not one in which the people generally were likely to have been much concerned; an abstract, colorless deity like Brahman could awake no sympathies in the hearts of those accustomed to worship gods of flesh and blood. Indeed, ever since the symbolical worship of nature had undergone a process of disintegration under the influence of metaphysical speculation, the real belief of the great body of the people had probably become more and more distinct from that of the priesthood.

In different localities the principal share of their affection may have been bestowed on one or another of the old gods who was thereby raised to the dignity of chief deity; or new forms and objects of belief may have sprung up with the intellectual growth of the people.

In some cases even the worship of the indigenous population could hardly have remained without exercising some influence in modifying the belief of the Aryan race. In this way a number of local deities would grow up, more or less distinct in name and characteristics from the gods of the Vedic pantheon. There is, indeed, sufficient evidence to show that, at a time when, after centuries of theological speculations, some little insight into the life and thought of the people is afforded by the literature handed down to us, such a diversity of worship did exist. Under these circumstances the policy which seems to have suggested itself to the priesthood, anxious to retain a firm hold in the minds of the people, was to recognize and incorporate into their system some of the most prominent objects of popular devotion, and thereby to establish a kind of creed for the whole community-subject to the Brahmanical law.

At the time of the original composition of the great epics two such deities, Shiva or Mahadeva (the great god) and Vishnu, seem to have been already admitted into the Brahmanical system, where they have ever since retained their place; and from the manner in which they are represented in those works, it would, indeed, appear that both, and especially the former, enjoyed an extensive worship. As several synonyms are attributed to each of them, it is not improbable that in some of these we have to recognize special names under which the people in different localities worshipped these gods, or deities of a similar nature which, by the agency of popular poetry, or in some other way, came to be combined with them. The places assigned to them in the pantheistic system were coordinate with that of Brahma; the three deities, Brahma, Vishnu and Shiva, were to represent a triple impersonation of the divinity, as manifesting itself respectively in the creation, preservation and destruction of the universe.

It is worthy of note that when the Vedic triad of Soma, Agni and Vayu was still recognized, attempts are made to identify Shiva with Agni,-in one passage in the Mahabharata it is stated that the Brahmins said that Agni was Shiva.

Although such attempts at an identification of the two gods remained isolated, they would at least seem to point to the fact that, in adapting their speculations to the actual state of popular worship, the Brahmans kept the older triad distinctly in view, and by means of it endeavoured to bring their new structure into harmony with the ancient Vedic belief.

As regards Vishnu, this god occupies already a place in the Vedic mythology, though by no means one of such prominence as would entitle him to that degree of exaltation implied in his character as one of the three hypostases of the divinity. This belief is not universally held as there are many Vedic verses that utter the oppositive view, *i.e.*, Vishnu's supremacy as a personal supreme God. Moreover, although in his general nature, as a benevolent, genial being, the Vedic god corresponds on the whole to the later Vishnu, the preserver of the world, the latter exhibits many important features for which we look in vain in his prototype, and which most likely resulted from sectarian worship or from an amalgamation with local deities.

In one or two of them, such as his names Vasudeva and Vaikuntha, an attempt may again be traced to identify Vishnu with Indra, who, as we have seen, was one of the Vedic triad of gods. The characteristic feature of the elder Vishnu is his measuring the world with his three strides, which are explained as denoting either the three stations of the sun at the time of rising, culminating and setting, or the triple manifestation of the luminous element, as the fire on earth, the lightning in the atmosphere and the sun in the heavens. This three strides corresponds with the events that took place when Vishnu incarnated as Vamana.

The male nature of the triad was supposed to require to be supplemented by each of the three gods being associated with a female energy (Shakti). Thus Sarasvati, the goddess of speech and learning, came to be regarded as the Shakti, or consort of Brahma; Sri or Lakshmi, beauty, fortune, as that of Vishnu; and Uma or Parvati, the daughter of Himavat, the god of the Himalaya mountain, as that of Siva. On the other hand, it is not improbable that Parvati, who has a variety of other names, such as Kali (the black one), Durga (the inaccessible one), Mahadevi (the great goddess) enjoyed already a somewhat extensive worship of her own, and that there may thus have been good reason for assigning to her a prominent place in the Brahmanical system.

A compromise was thus effected between the esoteric doctrine of the metaphysical and some of the most prevalent forms of popular worship, resulting in what was henceforth to constitute the orthodox system of belief of the Brahmanical community. Yet the Vedic pantheon could not be altogether discarded, forming part and parcel, as it did, of that sacred revelation (Sruti), which was looked upon as the divine source of all religious and social law, and being, moreover, the foundation of the sacrificial ceremonial on which the priestly authority so largely depended.

The existence of the old gods is, therefore, likewise recognized, but recognized in a very different way from that of the triple "revelation" of divinity. For while the triad represents the immediate manifestation of the eternal, infinite soul while it constitutes, in fact, the Parabrahman itself in its active relation to mundane and seemingly material occurrences, the old traditional gods are of this world, are individual spirits or portions of the Brahma-like men and other creatures, only higher in degree.

To them an intermediate sphere, the heaven of Indra (the svarloka or svarga), is assigned to which man may raise himself by fulfilling the holy ordinances; but they are subject to the same laws of being; they, like men, are liable to be born again in some lower state, and, therefore, like them, yearn for emancipation from the necessity of future individual existence. It is a sacred duty of man to worship these superior beings by invocations and sacrificial observances, as it is to honour the pitris (the fathers), the spirits of the departed ancestors.

Turiya: In early Hindu philosophy, turiya (also called caturtha) is a state of pure consciousness, or the experience of ultimate reality and truth. It is a fourth state of consciousness that underlies and at the same time transcends three common states of consciousness: (i) the state of waking consciousness (jagrata), (ii) the state of dreaming (svapna), (iii) and dreamless sleep (*susupti*).

Advaita Concept: The first two states are defective as experiences of reality and truth because of their flawed dualistic natures of subject and object, self and not-self, ego and non-ego. In the third state, dreamless sleep, one is not conscious of external or internal objects. But that does not mean consciousness is not present there. It is like saying 'I don't see anything in darkness'. The recognition that I don't see anything is what I 'see'. So also in dreamless sleep, one is not conscious of anything and the very fact that this statement is true proves the existence of consciousness during deep sleep.

In other words consciousness is the constant factor in all the three states and it is unaffected by the presence or absence of objects. Consciousness itself does not require to be revealed by another consciousness. It is self-revealed. While everything is presented to consciousness and is revealed by it, consciousness itself is not presented to anything else. It is never an object in relation to another subject. It is that which underlies both subject and object. It is the fourth, the turiya, the brahman. The Mandukya Upanishad defines turiya as follows:

> *"The fourth state is not that which is conscious of the subjective, nor that which is conscious of the objective, nor that which is conscious of both, nor that which is simple consciousness, nor that which is all-sentient mass, nor that which is all darkness. It is unseen, transcendent, the sole essence of the consciousness of self, the completion of the world."*

Four States of Consciousness: Turiya represents consciousness free from material influence. The idea is that consciousness, of which the atman is constituted, exists in our wakeful state of material experience, as it continues during sleep. In sleep we dream and experience the mental realm, whereas during our waking state the physical plane has more bearing on our lives.

Upon awakening from deep dreamless sleep one remembers existing in that condition. This is evidenced by the common expression, 'I slept well!' One cannot remember something one has no experience of. Thus in deep sleep when intelligence is transformed by tamo guna, the self continues to exist, as it does when intelligence is transformed by rajo guna during the dream condition and during the wakeful condition when intellect is transformed by sattva guna. The self is thus independent of the body and mind. If the physical and mental realms were to shut down, the self would continue to exist. This we know from our experience in deep sleep. Realizing this involves entering the turiya.

God as Turiya: Bhagavata Purana 11.15.16 describes Bhagavan with the words *turiyakhye* (the fourth), the meaning of which is found in the Bhagavad Gita 7.13:

tribhir guna-mayair bhavair ebhih sarvam idam jagat
mohitam nabhijanati mam ebhyah param avyayam
"Deluded by the three [gunas], the whole world does not know Me, who am above them and inexhaustible."

Fifth State in Gudiya Vedanta

The fourth dimension is the ground of our existence and the goal of all transcendentalists. For the Vedanta philosophers it is perceived variously, either as undifferentiated consciousness or a relationship with the divine. Regarding the latter, Gaudiya Vaishnava Theology concludes that love is greater than ourselves, and it is the greatest aspect of God, one that he himself is motivated by. For them, the nondual consciousness of Vedanta philosophy is realized when we know that we do not belong to ourselves, what to speak of anything belonging to us. If there is any time at which we can accurately say that something belongs to us, it is when, having given ourselves in love to God, we can say that "he is ours." Thus Gaudiya Vaishnavas are interested in *turyatitah gopala* (Lord Gopala beyond fourth dimension, Gopala-tapani Upanishad 2.96). This is the fifth dimension, in which one comes face to face with Gopala Krishna in Braj (Vraja Dhama), from adhoksaja to aprakrta, from God consciousness to Krishna consciousness. Jiva Gosvami elaborated on the turiya state in his Sandarbhas.

Guru-shishya Tradition

The guru-shishya tradition (also guru-shishya *parampara* or lineage) is a spiritual relationship found within traditional

Hinduism which is centered around the transmission of teachings from a *guru*. The term *shishya* roughly equates to the western term *disciple*, and in some parts of India is synonymous with the term *chela*. In a Hindu context, the term *guru* signifies one who teaches or imparts knowledge.

Such knowledge, whether it be Vedic, Agamic art, architecture, music or spirituality, is imparted through the developing relationship between guru and disciple. The principle of this relationship is that knowledge, especially subtle or advanced knowledge, is best conveyed through a strong human relationship based on ideals of the student's respect, commitment, devotion and obedience, and on personal instruction by which the student eventually masters the knowledge that the guru embodies.

The guru-shishya relationship is a practice which has evolved into a fundamental component of Hinduism, since the beginning of the oral traditions of the Upanishads (c. 2000 BC). The term Upanishad derives from the Sanskrit words *upa* (near), *ni* (down) and sad (to sit)—"sitting down near" a spiritual teacher to receive instruction in the guru-shishya tradition. An example of this dynamic can be found embodied in the relationship between Krishna and Arjuna in the Bhagavad Gita portion of the Mahabharata, and between Rama and Hanuman in the Ramayana. In the Upanishads, gurus and shishya appear in a variety of settings (husband answering questions about immortality, a teenage boy being taught by Yama, or Death personified, etc.). Sometimes the sages are women and at times the instructions (or rather inspiration) are sought by kings.

In the Vedas, the *brahmavidya* or knowledge of Brahman is communicated from guru to shishya by word of mouth, bringing about the traditional guru-shishya *parampara*, or lineage. The word Sikh is derived from the Sanskrit word shishya.

Common Characteristics of the Guru-shishya Relationship: Within the broad spectrum of the Hindu religion, the guru-shishya relationship can be found in numerous variant forms including Tantra. Some common elements in this relationship include:

- The establishment of a spiritual teacher/student relationship between the guru and shishya.
- A formal recognition of this relationship, which usually assumes the form of a somewhat structured initiation ceremony. In this process the guru both accepts the initiate as a shishya, and also agrees to assume some level of responsibility for the spiritual well-being and progress of the new shishya. During this initiation process a meditation mantra is often taught to the new shishya.
- Sometimes this initiation process will also include the teaching of other esoteric wisdom and/or meditation techniques.
- Gurudakshina-The *shishya* giving a valuable to the *guru* as a token of gratitude, often, apart from his services, the only monetary or otherwise fees that the student gives. Such tokens can be as simple as a fruit to as serious as a thumb as given by *Ekalavya* to his *guru Dronacharya*.

Guru-shishya Relationship Types: Amongst Guru-shishya relationships, there is a certain range of variation of the levels of authority that may be attributed to the guru. This level of authority ranges from the highest levels of authority as most often found in bhakti yoga, such as the Sathya Sai Baba movement, to the lowest levels, as found in the Pranayama forms of yoga such as in the Sankara Saranam movement. Between these two ends of the spectrum there are innumerable variations in degree and form of that authority.

Advaita Vedanta: Advaita vednta requires anyone seeking to study advaita vedanta to do so from a Guru (*teacher*). The Guru must have the following qualities (see Mundaka Upanishad 1:2.12):

1. Srotriya — must be learned in the Vedic scriptures and sampradaya

2. Brahmaniþha — literally meaning *established in Brahman*; must have *realised* the oneness of Brahman in everything and in himself

The seeker must serve the Guru and submit questions with all humility in order to remove all doubts (see Bhagavad Gita 4.34). By doing so, advaita says, the seeker will attain moksha (*liberation from the cycle of births and deaths*).

Sruti Tradition

The Guru-shishya tradition plays an important part in the continuation of the Shruti tradition in Vaidika dharma. The Hindus believe that Vedas are handed down through the ages from Guru to shishya. The Vedas prescribe for a young brahmachari to be sent to a Gurukul where the Guru (referred to also as acharya) teaches the pupil the Vedas as also the Vedangas. The pupil is also taught the prayoga to perform yajnas. The term of stay varies (Manu Smriti says the term may be 12 years, 36 years or 48 years). After the stay at the Gurukul the brahmachari returns home after performing a ceremony called samavartana.

The word Srauta is derived from the word Sruti meaning that which is heard. The Srauta tradition of handing the Vedas down the generations consisted in solely of oral tradition from Guru (teacher) to the Sishya (student). But of late many Vedic scholars make use of books in order to teach the Vedas to their students.

Bhakti Yoga

The most well known form of the Guru-shishya relationships is the bhakti guru-shishya relationship. Bhakti (Sanskrit) "Devotion" means surrender to God, gods or guru (sanskrit *guru bhakti* = devotion to the teacher). Bhakti extends from the simplest expression of devotion to the ego-decimating principle of prapati, which is total surrender. The bhakti form of the guru-shishya relationship generally incorporates three primary beliefs or practices:

1. Devotion to the guru as a divine figure or avatar

2. The belief that such a guru has transmitted, or will eventually impart moksha, diksha or shaktipat to the (successful) shishya
3. The belief that if the shishya's act of focusing his or her devotion (bhakti) upon the guru is sufficiently strong and worthy, then some form of spiritual merit will be gained by the shishya

Within the bhakti form of the guru-shishya relationship, a certain type of fixed dependency sometimes develops between the guru and the shishya, that may in some ways be similar to the relationship between a parent and a child.

Prapati: In the ego-decimating principle of prapati (Sanskrit, "Throwing oneself down"), the level of the submission of the will of the shishya to the will of the guru is sometimes extreme. It is one of total, unconditional submission to God or guru, often coupled with the attitude of personal helplessness, self-effacement and resignation. This doctrine is perhaps best expressed in the teachings of the four Samayacharya saints, who all shared a profound and mystical love of Siva that included:

- Deep humility and self-effacement, admission of sin and weakness
- Total surrender in God as the only true refuge and
- A relationship of lover and beloved known as bridal mysticism, in which the devotee is the bride and Siva the bridegroom

In its most Extreme form it Sometimes includes:

- The assignment of all or many of the material possessions of the shishya to the guru
- The strict and unconditional adherence by the shishya to all of the commands of the guru. An example is the legend that *Karna* silently bore the pain of a wasp drilling into his flesh on the thigh so as not to disturb his *guru Parashurama*.
- A system of various titles of implied superiority or deification which the guru assumes, and often requires the shishya to use whenever addressing the guru

- The requirement that the shishya engage in various forms of physical demonstrations of affection towards the guru, such as bowing, kissing the hands or feet of the guru, and sometimes agreeing to various physical punishments as may sometimes be ordered by the guru
- Sometimes the authority of the guru will extend to all aspects of the shishya's life, including sexuality, livelihood, social life, etc.

In exchange for such absolute submission to the direction of the guru, the shishya usually expects the guru to provide the spiritual guidance necessary for the shishya to achieve some sort of meaningful spiritual progress. Often a guru will assert that he or she is capable of leading a shishya directly to the highest possible state of spirituality or consciousness, sometimes referred to within Hinduism as moksha. In the bhakti guru-shishya relationship the guru is often believed to have supernatural powers, which would be consistent with the deification of the guru that is usually a part of this relationship.

Other: Even though bhakti yoga has traditionally been practiced by the large majority of Hindus, a small minority spiritual teachers have for some time advocated non-bhakti forms of the guru-shishya relationship. Amongst these is Lahiri Mahasaya who refused many of the popular devotions proffered to him by those who wished to *deify* him, and who retained his regular job as a clerk, living as a regular householder, meanwhile being acclaimed as a great spiritual teacher by many of his shishyas. Lahiri also refused to allow his shishyas to organize any kind of a religious institution centered upon himself.

Also amongst these is Sankara Saranam, a former shishya of Paramhansa Yogananda. Saranam insists that the guru may only act as a teacher that is more akin to a common academic teacher. As such, it is only the shishya who can accomplish his or her own spiritual advancement via his or her own learned ability to look within. As such, just as within academics, it is ultimately only the student who can actually advance him or herself by his or her own efforts, while such advancement is merely enhanced by the fertile learning environment that is created by the worthy teachings and insightful observations provided by a suitable guru. Some elements that are common to most non-bhakti guru-shishya relationships are:

1. The guru is self-supporting, and does not accept supporting donations from shishyas
2. The guru is unwilling to assume responsibility for, or give specific advice regarding the life decisions of the shishya
3. The guru insists on a fundamental equality with his or her shishyas, and refuses to accept any honorific titles from his or her shishyas

Buddhism: In the Theravada Buddhist tradition, the teacher is a valued and honoured mentor worthy of great respect and a source of inspiration on the path to Enlightenment. In the Tibetan tradition, however, the teacher is viewed as the very root of spiritual realization and the basis of the entire path. Without the teacher, it is asserted, there can be no experience and insight.

The guru is to be seen as Buddha. In Tibetan texts, great emphasis is placed upon praising the virtues of the guru. Tantric teachings include generating visualisations of the guru and making offerings praising the guru. The guru becomes known as the *vajra* (literally "diamond") guru, the one who is the source of initiation into the tantric deity. The disciple is asked to enter into a series of vows and commitments that ensure the maintenance of the spiritual link with the understanding that to break this link is a serious downfall.

Psychological Aspects: Rob Preece, in *The Noble Imperfection*, writes that while the teacher/disciple relationship can be an invaluable and fruitful experience, the process of relating to spiritual teachers also has its hazards. These are the result of naivete amongst Westerners as to the nature of the guru/devotee relationship and the consequence of a lack of understanding on the part of Eastern teachers as to the nature of Western psychological makeup. Preece introduces the notion of transference to explain the manner in which the guru/disciple relationship develops from a more Western psychological perspective. He writes: "In its simplest sense transference occurs when unconsciously a person endows another with an attribute that actually is projected from within themselves." In developing this concept, Preece writes that when we transfer an inner quality onto another person we may be giving that person a power over us as a consequence of the projection, carrying the potential for great insight and inspiration, but also the potential for great danger. "In giving this power over to someone else they have a certain hold and influence over us it is hard to resist, while we become enthralled or spellbound by the power of the archetype".

Ritual

A ritual is a set of actions, performed mainly for their symbolic value, which is prescribed by a religion or by the traditions of a community.

A ritual may be performed at regular intervals, or on specific occasions, or at the discretion of individuals or communities. It may be performed by a single individual, by a group, or by the entire community; in arbitrary places, or in places especially reserved for it; either in public, in private, or before specific people. A ritual may be restricted to a certain subset of the community, and may enable or underscore the passage between religious or social states.

The purposes of rituals are varied; they include compliance with religious obligations or ideals, satisfaction of spiritual or emotional needs of the practitioners, strengthening of social bonds, demonstration of respect or submission, stating one's affiliation, obtaining social acceptance or approval for some event — or, sometimes, just for the pleasure of the ritual itself.

Rituals of various kinds are a feature of almost all known human societies, past or present. They include not only the various worship rites and sacraments of organized cults and religions, but also the rites of passage of certain societies, oaths of allegiance, coronations, and presidential inaugurations, marriages and funerals, school "rush" traditions and graduations, club meetings, sports events, halloween parties and veteran parades, Christmas shopping, and more. Many activities that are ostensibly performed for concrete purposes, such as jury trials, execution of criminals, and scientific symposia, are loaded with purely symbolic actions prescribed by regulations or tradition, and thus partly ritual in nature.

Even trivial actions like hand-shaking and saying hello are rituals. In any case, an essential feature of a ritual is that the actions and their symbolism are not arbitrarily chosen by the performers, nor dictated by logic or necessity, but are, at least in part, prescribed and imposed upon the performers by some external source.

Ritual Actions: Because of their symbolic nature, there are hardly any limits to the kind of actions that may be incorporated in a ritual. The rites of past and present societies have typically involved special gestures and words, recitation of fixed texts, performance of special music, songs or dances, processions, manipulation of certain objects, use of special dresses, consumption of special food, drink, or drugs, and much more. Religious rituals have also included animal sacrifice, human sacrifice or ritual suicide, and other forms of ritual murder.

Purposes: Ritual serves diverse purposes including:

- Worship
- Ritual purification with the aim of removing uncleanliness, which may be real or symbolic.
- Atonement
- Dedication

Religious Ritual: In religion, a ritual can comprise the prescribed outward forms of performing, the *cults* or cult of a particular observation within a religion or religious denomination. Although ritual is often used in context with worship performed in a church, the actual relationship between any religion's doctrine and its ritual(s) can vary considerably from organized religion to non-institutionalized spirituality, such as ayahuasca shamanism as practiced by the Urarina of the upper Amazon. Rituals often have a close connection with reverence, thus a ritual in many cases expresses reverence for a deity or idealized state of humanity.

Sociology

Outside worship and reverence, rituals can have a more basic sociological function in expressing, fixing and reinforcing the shared values and beliefs of a society. Ritual thus is as much political as religious in orientation. Rituals have formed a part of human culture for tens of thousands of years. The earliest known evidence of burial rituals dates from around 20,000 years ago. (Older skeletons show no signs of deliberate 'burial', and as such lack clear evidence of the materialization of ritual.) Rituals can aid in creating a firm sense of group identity. Humans have used rituals to create a social bond and alleviate the isolation that can be felt otherwise.

Fraternal: Nearly all fraternities and sororities have secret rituals incorporated into their structure. Many were founded in college and university societies resistant to the idea of a fraternal organization. Thus, numerous aspects of ritual and ritualistic proceedings are ingrained into the workings of the societies.

Psychology: In psychology, the term *ritual* sometimes refers to a specific action or series of actions that a person performs in a given context which otherwise has no apparent reason or purpose. The term may refer especially to compulsive behaviors of people afflicted with obsessive compulsive disorder (OCD).

Jyotisha: Jyotisha (*jyotica*, in Hindi and English usage Jyotish) is the Hindu system of astrology, one of the six disciplines of Vedanga. The Sanskrit word derives from *jyótis* (disjointed as "Ja"+"Ya"+"O"+"T"+"ish" to get the root meaning as "water or birth" + in addition to" + "earth" and "stars" + "knowledgeable" equating as one who is knowledgeable, or enlightened with knowledge, of birth, fate, and relationship to water, earth and stars) or which means "light, brightness", but in the plural also "the heavenly bodies, planets and stars". *Jyotish* may thus at one level be seen to signify a "science of heavenly bodies". However, as the singular purpose of jyotish and the vedas is to dispel the darkness of ignorance, it is more commonly taken to signify the "science of light". Indeed, Jyotish has historically been part of a continuous "holistic" approach to living and to spiritual practice within the life of Hindus predominant in India. Jyotish is often discussed as the instructional element of the Rig Veda, and as such is a Vedanga, or "body part" of the Vedas. Jyotish is called the *Eye of the Veda*, for its believed ability to view both phenomenal reality and wisdom itself.

Part of a larger Vedic curriculum including mathematics, architecture, medical and military applications, it superficially has much in common with ancient and modern Western astrology (and the early traces of the descent of various schools of astrology from the Harappan and Egyptian cultures, Chinese and the Chaldean, through the Arabs, Greeks, and early Romans show complex interweavings that are assessed variously by diverse camps of scholars — who, however, currently find little common ground as to the exact historical development). Jyotish has many facets, and some of its basics are clearly also cornerstones of Western astrology, such as symbolically endowed signs, houses and planets. But Jyotish has its own sophisticated reference to the nominal: the planets are "grahas", forces that seize or act upon created beings and influence their actions and life. As Hindus believe that humans have ups and downs in life because of karma, many use Jyotish to understand the downs in life due to the influence of planets, *i.e.*, Navagraha and perform religious ceremonies to mitigate the bad karma.

Foundations: Jyotish's many lineages or paramparas emphasize that its study is a sadhana or technique of mental and existential development. In modern times it is a chief source of reference for many Hindus and other spiritual practitioners. In addition to sustaining its own lore, commentary and debates on the dilemmas of fate and free will and the philosophy of karma, jyotish maintains a virtual *materia medica* of remedies for difficulties found in horoscopes. Vedic astrologers will frequently prescribe special stones, or specific therapeutic practices or meditation techniques using mantras to those facing difficult or unclear futures as predicted by means consistent with Jyotish methodology. In past centuries, Brahmins have been the primary practitioners of Jyotish. In the last century, a renaissance of study of Jyotish and other Vedic sciences emerged in India and the west.

The foundation of jyotish is the notion of bandhu of the Vedas, which is the connection between the microcosm and the macrocosm. Surya, the Sun, is a manifestation of Vishnu, a central aspect of the Supreme, and is also the Atman within man. The term "Vedic astrology" has been recently introduced by American and Western astrologers in the 1980s and 1990s, leading to collaborative organizations such as the now-international Council of Vedic Astrology. Jyotish is still commonly used to aid in important decisions in modern India. In Hindu culture, newborns are traditionally named based on their jyotish charts, and jyotish concepts are pervasive in the organization of the calendar and holidays as well as in many areas of life and science. Astrology is vital to decisions made about marriage, opening a new business, and moving into a new home.

Branches of Vedic/Hindu Astrology

Vedic astrology (*Aagam Shaastra / Jyotisha*-futurity sciences) has three main branches:

- *Siddhanta (Astronomy):* Astronomy & its application to astrology
- *Samhita (Mundane astrology):* Covers Mundane astrology, predicting important events related to countries such as war, earth quakes, political events, astro-meteorology, financial positions, electional astrology; house & construction related matters (Vaastu Shaastra), animals, portents & omens etc.

Hora (Predictive Astrology): This branch has the following different styles/subbranches:

- *Jaatak Shaastra*/Hora Shaastra (Natal Astrology/ horoscopy): Prediction based on individual horoscope.
- *Muhurt or Muhurtha* (Electional astrology): Selection of beneficial time to initiate an activity to get maximum fruition from the life activities.
- *Swar Shaastra* (Phonetical astrology): Predictions based on name & sounds.

- *Prashna* (Horary astrology): Predictions based on time when a question is asked by querent / querist.
- *Ankjyotisha / Kabala* (Numerology): A branch of astrology based on numbers.
- *Nadi* Astrology: An ancient treatise having detailed predictions for individuals.
- *Tajik Shaastra / Varsha Phal* (Annual Horoscopy): Astrology based on annual solar returns.
- *Jaimini Sutras*: A non-conventional method of timing of events based on Famous Indian astrologer, Acharya Jaimini.
- *Nastjaatakam* (Lost Horoscopy): Art of tracing / construction of lost horoscopes.
- *Streejaatak* (female astrology): A special branch of astrology dealing with female nativities.

Other Related Branches

- *Graha Samudriki* (Astro-Palmistry): Palm reading as horoscope.
- *Hasta Rekha / Samudrika Shaastra* (Palmistry): Based on palm reading.
- *Padatal Shaastra* (Plantarology): Based on reading of lines & signs on the sole.
- *Shakun Shaastra* (Omens): Predictions based on omens & portents.
- *Swapna Vidhya* : Interpretation of dreams.
- *Kapal Vidya* (Phrenology)
- *Aakriti Vidya* (Physiognomy): Based on structure & moles on the body.
- *Kerala Jyotisha*: Predictions based on querrist reply regarding name of flower or colour or touching part of body.
- Remedial Astrology: Various modes of propitiation of planets based on planetary positions in nativity, transits, elections & for religious functions.

In Vedic literature guidance is given on two other branches viz, Electional astrology & Remedial astrology.

2

Buddhist Philosophy

A singing bowl, also known as a Himalayan bowl or cup gong, is a musical instrument used in Buddhist meditation, dating back many centuries. They are traditionally played in Tibet, Nepal, India, China, Japan, and Korea, mostly by Buddhist monks and in monasteries.

The instrument is made of metal, often including precious metals such as silver and gold. Many other metals, including copper, tin, zinc, lead, iron, and, nickel also feature in many bowls' metallic makeup. Generally, singing bowls can be made of many different metals.

Singing bowls are played by hitting, striking, or slowly rubbing/stirring a wooden striker/mallet within or on the top outer surface of the bowl. When done correctly, the bowl will produce a continuous harmonic sound, said to be relaxing and good for the chakra. Some claim that many bowls are "tuned" to a specific chakra. Currently, these bowls are used for meditation, yoga, healing, sound massage, sound therapy, and chakra balancing.

Singing bowls can be found in a variety of shapes, sizes, designs, colours, carvings and craftings. They may be found with various religious themes and symbols, like Tibetan mantra chant om mani padme hum, dragon carvings, mandala carvings, *asta mangal* (eight auspicious good luck symbols), Buddhist religious symbols and many more. The making of traditional bowls is now considered a lost art. Most old bowls contain very little, if any, ornamentation.

Most bowls are handmade, apart from some tools used for shining and polishing, and fall into two categories. The first is the hand-hammered (beaten) singing bowl, most of which come from Nepal and India, and are manufactured using traditional processes. Hand-hammered bowls are very much more popular for meditation, healing and sound therapy. The second type is the colored, carved and crafted bowl. These are popular for gifts, decorations and souvenir items, and are mostly from Nepal, the Himalayan Country.

Basic Buddhist Philosophical Doctrines: There are some philosophical doctrines that are so early and so fundamental to Buddhism that denials of them tend to be regarded as profoundly non-Buddhist heterodoxies. All forms of Buddhism endeavor to maintain these principles.

1. *Momentariness:* Nothing exists for any length of time. There is no *substance* or duration to things. Each moment is an entirely new existence, which is succeeded by an entirely new existence. The only connection between one thing and the next is that one causes the next. This doctrine sounds much like the Greek philosopher Heraclitus. The "things" tend to be called the *dharmas* in Buddhist thought.
2. *Relative Existence or No Self Nature:* Nothing has a *essence*, nature, or character by itself. Things in isolation are *shunya*, "empty." The nature of things only exists in relation to everything else that exists. Existence as we know it is thus completely relative and conditioned by everything else. Only *Nirvana* would be unconditioned, although we cannot know what it is like. The distinction between the conditioned reality that we know and the unconditioned reality that we do not know is similar to the distinction between *phenomena* and *things-in-themselves* made by the German philosopher Immanuel Kant. The notion that the *dharmas* derive their nature from everything else has led to comparison with the "monads" of another German philosopher, Leibniz. The monads also represent the whole universe. However, since the

dharmas are momentary, this is actually more like the actual entities postulated by Alfred North Whitehead (1861-1947) in his "Process" philosophy. *Shunyata*, "Emptiness," is easily misunderstood. It is not nothingness. Emptiness is neither existence, nor non-existence, nor both existence and non-existence, nor neither existence nor non-existence. At the very least, this means that we don't know what is left when we take away all conditioned relations. Beyond that, it can mean that we *cannot* know what that is. No Self Nature means that there are no *essences*, just as Momentariness means that there are no *substances*.

3. *No-Atman:* There is no Self (*atman*) in Buddhism, either as an essence or as a substance. What we call our self is a collection of things, the "aggregates" (*skandhas*): 1) the body, or "form," 2) feelings, 3) ideas, 4) impressions, & 5) momentary consciousness. There is no enduring thing present in the aggregates. This critique of the self as just a collection is very similar to the view of the Scottish philosopher David Hume. An implication of No-Atman is that reincarnation cannot be *transmigration*, since there is nothing to migrate.
4. *No-God:* There is no Brahman or any other such ultimate enduring substance or nature to reality. *Nirvana* thus cannot be characterized as realizing either Self, Brahman, or God.
5. *Dependent Origination:* Everything has a cause. A momentary existence occurs as it does because of a previous momentary existence, but the cause itself is also momentary. Dependent Origination combines the doctrines of momentariness and relative existence and is why in the Second Noble Truth desire and ignorance cause each other. That relationship can be expanded:

1. ignorance (*avidya*), causes
2. impressions (*samakaras*), which cause
3. consciousness (*vijnana*), which causes
4. mind-body (*namarupa*), which causes

5. the sense organs (*sadayatana*), which cause
6. contact with objects (*sparsha*), which cause
7. experience (*vedana*), which causes
8. desire, which causes
9. clinging (*upadana*), which causes
10. the will to be born (*bhava*), which causes
11. rebirth (*jati*), which causes
12. suffering (*jaramarana*), which in turn causes
13. ignorance (*avidya*).

Nirvana is thus not the removal of an ultimate cause but the simultaneous removal of *all* causes, all of conditioned existence.

6. *Karma:* Because there is no substance or duration in Buddhism, the Buddhist view of karma is different from that in Hinduism or Jainism. Karma is only causation, without the mediation of any substance (*apurva*, causal body, etc.). Reincarnation thus consists in our being *caused* by something in the past, and our karma is simply the effect now of past actions.

In the history of Buddhist philosophy, these doctrines created some difficulties. If there is no self, then what is it that attains enlightenment or Nirvana? It is not me, for I am already gone in an instant; and if it is not me, then why bother? Also, if there is no enduring self, then the rewards and punishments of karma are visited on different beings than those who merited them. Why do I, instead of someone else, deserve the karma of some past existence? The Buddha himself probably would have been irritated with the doctrines that created these difficulties, since he rejected theorizing (it did not "tend to edification"), and he would have expected no less than that such theories would lead to tangled and merely theoretical disputes.

The important philosophical lesson of these difficulties, however, is whether the concept of causality can be used as a substitute for the concept of substance. In all honesty, no. Something rather like the Buddhist position, however, can be formulated by Kant, for whom the concept of substance applies to phenomena but has only uncertain meaning when applied to things-in-themselves. Phenomena are only "provisional existence" to Buddhism, and the Buddhist doctrine of no enduring Self could easily be adapted to the Kantian transcendent.

Stages in the History of Buddhism: The history of Buddhism in India, which lasted about 1500 years, can be divided into 500 year periods, during which distinctive forms of Buddhism emerged. This is an idealized and schematic picture, but it is convenient, and it can be matched up with where Buddhism spread during these periods and what forms of Buddhism became dominant there.

Buddhist doctrine and practice in the earliest period were agreed upon in a series of Councils, sometimes reckoned to be three, or four.

1. The First Council was held shortly after the Buddha's death, at Rajagriha. Issues about the conduct of the Sangha, the monastic community, in the absence of the Buddha appear to have been settled.
2. The Second Council was held about a century after the Buddha's death, at Vaishali. Sometimes this is considered the First Council, or is confused with the previous one. It began to agree on the content of the Buddhist Canon and on the monastic discipline, the vinaya.
3. The Third Council was called by the Emperor Ashoka and held at Pataliputra. The content of the Pali Tripitaka, "Three Baskets," is supposed to have been settled at this Council, or there abouts in this period, consisting of the Sutra-pitaka [Sutta in Pali], the sermons of the Buddha, the Vinaya-pitaka, the monastic discipline, and the Abhidharma-pitaka [dharma is dhamma in Pali], the philosophical development of Buddhist doctrine. Buddhism was introduced into Sri Lanka (Ceylon) as the result of Ashoka's own efforts, and the Canon in Pali was preserved there.

4. The Fourth Council was called under the Emperor Kanishka I and held at Jalandhara. This is not attested in Pali sources, and so one often hears that there were only three Councils. The Council is supposed to have supervised the translation of the Tripitaka into Sanskrit. The Canon apparently had not only existed in Pali, but in other Prakrits, which were all consulted for a standard Sanskrit version. It appears to be the Sanskrit texts that were subsequently spread to China. The sutras of the Mahayana may have existed only in Sanskrit from the beginning.

Theravada (Teaching of the Elders) Buddhism (called "Hinayana," the "Lesser Vehicle," by the Mahayana): In India, 5th century BC to 1st century AD.

1. Distinctive doctrines:
2. The Buddha (Siddhartha Gautama, Shakyamuni) is gone, and individual practitioners must work out their salvation on their own.
3. The Buddha was unique, and individual practitioners cannot become Buddhas, only *arhats* (saints). There will be a future Buddha, Maitreya, but not for thousands of years.
4. *Nirvana* (liberation) and *samsara* (the place of death and rebirth) are definitely different. *Samsara* is a place of suffering to be left behind. *Nirvana* is a liberation that is free of death and rebirth but is beyond description and rational understanding.

Mahayana (Great Vehicle) Buddhism: In India, 1st century AD to 6th century. is not gone, and individual practitioners are not on their own. Instead, the Buddha taught the dharma out of compassion, and his compassion would prevent him from being unavailable to practitioners now. Indeed, to emulate the compassion of the Buddha, practitioners become bodhisattvas, who vow to carry all beings with them into salvation. Bodhisattvas are also available, like the Buddhas, to help people work out their salvation. Maitreya is presently a bodhisattva, but the most important bodhisattva is probably

Avalokiteshvara, who became identified with the Chinese goddess of Mercy, Guanyin (Kwan-in in Wade-Giles, Kannon in Japan).

The Buddha was not unique, and individual practitioners who have become bodhisattvas can become Buddhas. There are already multiple Buddhas besides Shakyamuni. Most important are Mahavairocana and Amitabha. Amitabha is famous for his Western Paradise, or Pure Land, where he has Vowed to cause anyone who calls on him for help to be born, so they will be free of the world of suffering to work out their ultimate liberation. In Japan Amitabha is known as Amida and Mahavairocana as Dainichi. Most of the famous Buddha statues in Japan are not Shakyamuni: the great outdoor bronze Buddha at Kamakura is Amida, and the Buddha enshrined in the Todaiji ("Great Eastern") Temple in Nara (the largest wooden building in the world), is another Buddha named Locana.

Nirvana and samsara are no longer definitely different. The "Fourfold Negation" is applied to the relationship between the two. Samsara and Nirvana are thus neither the same, nor different, nor both the same and different, nor neither the same nor different. This allows some room for maneuver, which may have made Buddhism more palatable in China, where Confucianism never did approve either of the world-denying metaphysics or the monasticism of Buddhism. Distinctively Chinese schools of Buddhism developed, like T'ien-t'ai (Tendai in Japan) and Ch'an (Seon [Son] in Korea, Thien in Vietnam, Zen in Japan), for whom samsara and Nirvana were virtually identical, so that enlightenment and Nirvana transformed the world rather than eliminated it.

The paradoxical metaphysics of Buddhism could be assimilated to the similar paradoxical doctrines of the native Chinese philosophical school of Taoism.

- *Places where Mahayana spread:* Mahayana Buddhism is presently practiced in China, Korea, Vietnam, and Japan. Buddhism was propagated in China by missionaries from India, like Kumarajiva (344-413), who arrived in China in 401, and Buddhabhadra (359-

429), who arrived in 408, and by Chinese pilgrims who traveled to India, like Fa-Hsien (Fa³xian³), who travelled to India between 399 and 414, and Hsüan-tsang (Xuánzang³, 600-664), who went to India between 629 and 645. These were difficult journeys, either by sea around Malaya, where many ships were attacked by pirates or sunk by storms, or by land through Central Asia, on the "Silk Road" caravan route, through deserts and over some of the highest mountains in the world. The highest peaks of the Pamirs and related mountain chains, called Bam-e Donya in Persian, "the Roof of the World," are all over 24,000 feet [note]. One story of a Buddhist missionary crossing the Pamirs, Kumarayana, father of Kumarajiva, is that he carried a Buddha image during the day — and the Buddha image carried him during the night! Kumarajiva, Fa-Hsien, and Hsuan-tsang all brought Buddhist texts from India to China and translated them.

The Buddhist canon as it arrived in China was in Sanskrit, and it included many special Mahayana Sutras that are not in the Pali Canon (though many are now suspected of being Chinese forgeries). The stories of Fa-Hsien and Hsüan-tsang's travels are important parts of Chinese literature, and Fa-Hsien's account of India during the reign of Chandra Gupta II (375-415) is an important document for the history of India.

Vajrayana ("Thunderbolt Vehicle") Buddhism: In India, 6th to 11th century

- *Distinctive doctrines:* Vajrayana Buddhism is Tantric Buddhism, often called "esoteric" Buddhism. Although it is sometimes also translated as "diamond" (*i.e.* "hard"), the vajra (kongo in Japanese) was originally the thunderbolt of Indra; and in Vajrayana it symbolizes the magical power of Tantrism. Tantric magic could be worked through mandalas, sacred diagrams, mantras, sacred formulas for recitation (the most famous one being, "Om, mane padme hum" — "The jewel is in the lotus") and mudras, sacred gestures. This Tantric magic could be merely thaumaturgical ("wonder working") or could be regarded as means of achieving liberation in addition to or apart from meditative or meritorious practices. Just as Hindu Tantrism expresses its magical power through goddesses like Kali, Vajrayana emphasizes female figures. Vajrayana comes to balance male Bodhisattvas with female Bodhisattvas as attendants of the various Buddhas. And while Buddhas tend to be regarded as male in all branches of Buddhism, Tibetan Buddhism supplies female figures corresponding to each Buddha, like the "savioresses" Green Tara, White Tara, and Mamaki, who actually vow to always be reborn as women in the process of leading all beings to salvation. Vajrayana symbolism always balances male and female: the Vajra Mandala (or the "jewel" above) corresponds to the Womb (or Matrix) Mandala (the lotus). The extent to which Vajrayana practiced real sexual union, been the physical male "vajra" and the physical female womb, as part of its Tantrism is unclear and disputed. Often "right-handed."

Tantrism is distinguished from "left-handed" Tantrism, in which the former practiced the union of male and female, in symbolic, iconographic form, while the latter practiced it literally. While the "right-handed" forms are mainly what remain in Tibet and in Japanese Shingon today, there is little doubt that real "left-handed" practices existed in the past and survive to an extent in the present, and Tibetan art sometimes still portrays the more violent and disturbing aspects of Tantric practice — rape, bestiality, etc.

- *Places where Vajrayana Spread:* Vajrayana Buddhism most importantly spread to Tibet and then Mongolia. In Tibet it assumed distinctive forms that are usually called Lamaism, since the monks are called Lamas. The present Dalai Lama, who was the priestly ruler of Tibet until he fled the Communist Chinese in 1959, is from a line that is reputed to be successive incarnations of the Bodhisattva Avalokiteshvara.

Vajrayana Buddhism also entered China, Japan, etc. as special "esoteric" schools, like the Japanese Shingon school. The great temple at Borobudur, outside Jakarta on the island of Java, dates from this period (c. 800) and embodies Vajrayana Mandala forms; but in Indonesia Buddhism soon thereafter gave way to Islam.

The End of Buddhism in India: Buddhism may have died out in India in the 11th century because:

1. It had become almost indistinguishable from the Tantric forms of Hinduism. Sophisticated Buddhist doctrine did not appeal to most people, and the actual practices and iconography of Vajrayana could easily be assimilated into Hinduism. And,
2. Islam arrived in earnest in India with the Afghan prince Mahmud of Ghazna, who defeated a coalition of Hindu princes in 1008 and soon annexed the Punjab. As Buddhism was persecuted, conversions to Islam increased, and Buddhism declined. By the time the British arrived, about 25% of India was Moslem. That ultimately led to the partition of the country into India and Pakistan. The Gautama Buddha himself has ended up being regarded as the 9th Incarnation (Avatar) of the great Hindu God Visnu, although the unflattering take on it is that he deliberately taught a false doctrine (*i.e.* Buddhism) in order to deceive and destroy demons.

The Final Dharma Age: Another way of dividing the history of Buddhism emerged in the Buddhist tradition as a way of dealing with the prediction of the Buddha himself that the Dharma would only last 500 years. This became a matter of concern in China, where Buddhism did not even become established until nearly 1000 years after the time of the Buddha. Indeed, there was uncertainty about when the time of the Buddha had been, but soon enough it was obvious that far more than 500 years had passed. The doctrine that was formulated in response to this we find in K'uei-chi (Tz'u-en, 632-682), founder of the Fa-hsian school early in the T'ang Dynasty, in his I-lin-chang, "The Grove of Meanings." There we find a division into three periods based on the existence of Buddhist teaching, Buddhist practice and Buddhist "proof," *i.e.* results—supernatural powers and Nirvana:

- The Age of the True Dharma (Sanskrit Saddharma, Japanese shobo). In the first 500 years of Buddhism (sometimes put at 1000 years, however), the teaching, the practice, and the proof of Buddhism are all evident and effective. This becomes the Buddhist equivalent of the "age of the Apostles" in Christianity or of the "companions of the Prophet" in Islam — when the best exemplars of the religion lived, people whose achievements are now largely beyond the reach of believers (as most Protestants do not believe that any modern Christians can perform miracles on the scale of the Apostles). It becomes, naturally, particularly associated with Buddhism in India, the first step in the Sangoku, or "three nations," ideology that later developed in Japan.
- The Age of the Counterfeit Dharma (Sanskrit Saddharma-pratirupaka, Japanese zoho). In the 1000 years following the True Dharma, the teaching and the practice of Buddhism are evident but the proof fails. While uncertainty about the length of the True Dharma Age could postpone a reckoning, Buddhism in China — which now becomes particularly associated with the second Age — eventually had to face the fact that the full force of the Dharma, as defined by the Buddha himself, was spent. This would seem to remove the point of Buddhist belief and practice altogether, but it was also something whose full implications most Buddhists were unwilling to accept. One result was the popularity of devotionalistic forms of Buddhism, which substitute intermediate goals before final Nirvana. In Pure Land practice, the Vow of the Buddha Amitabha enables devotees to be reborn in a Pure Land, where they will be free of suffering, let alone the Hells, and can work out their salvation without distractions (like hunger or sex). Pure Land Buddhism, indeed, became the most popular form of Buddhism in China and Japan.

Another approach, which we see in the T'ien-tai (Japanese Tendai) and Ch'an (Zen) Schools, was to redefine the "proof," so that avoiding rebirth was no longer the goal of Buddhist practice. Instead, life in the world of samsara can become the equivalent of Nirvana. This gives to Buddhism itself a very different character, unlike the world-denying tendencies of India and more conformable to the sensibilities of both Confucianism and Taoism. This goes along with discussions about whether it was really necessary to keep Buddhist Precepts. Nevertheless, we do have a continuation of traditional monasticism, and of course not everyone agrees with the new tendencies in thinking.

- The Age of the Final Dharma (Sanskrit Saddharma-vipralopa, Japanese mappo). If the Counterfeit Dharma posed a disturbing challenge for Buddhism, the Final Dharma was even worse. At that point the real practice and proof of Buddhism are both supposed to no longer be possible. Only the teaching remains. While K'uei-chi didn't yet quite have to worry about that, by the time we get to the Kamakura Period (1186-1336) in Japanese history, it becomes widely accepted that the Final Dharma, Mappo, is upon them. This is expected to last no less than 10,000 years. After that, Buddhism will simply be forgotten, and ages will pass, perhaps a million years, before the arrival of the Future Buddha, Maitreya. By this time, of course, Buddhism was already vanishing in India itself. Some Japanese expected this, or were aware of it, others didn't believe it. But it was the kind of thing to reinforce the sense of the fading of the Dharma.

The response was a period of religious creativity, in which a number of Chinese schools were adopted and reworked (Tendai, Zen, Jodo), and some original Japanese schools (Jodo Shin, Nichiren) emerged. Japanese monks ceased taking full monastic vows, clerical marriage became tolerated, sometimes sanctioned (in Jodo Shin), later even expected, and Japanese monks could easily gain reputations as formidable drinkers, despite the Precept against intoxicants. To Buddhists elsewhere, this kind of thing often makes it seem that Japanese monks are no longer proper Buddhists. In the Japanese tradition, however, this is all of a piece with Mappo, and the development of Japanese Buddhism was seen as actually making it the living center of the Buddhist world.

Thus, in the Sangoku ideology, Japan itself becomes the exemplar of the third Age. This would have some unfortunate implications, when Imperial Japan saw itself as exporting its superiority to the rest of Asia. Japanese Buddhist "missions" in Korea helped discredit Buddhism to many, resulting in large numbers of Koreans converting to Christianity. Nevertheless, Japanese religious creativity remains impressive, as groups of "New Religions" continue to draw on Buddhist and various eclectic sources, and some schools, like Nichiren, have managed to establish themselves among non-Japanese around the world.

While many Buddhists now no longer worry about the problem of the fading Dharma, there is no denying the statement of the Buddha, or the role that dealing with this has played in the history of Buddhism.

Culmen Mundi, the Roof of the World: Most people think of the Himalayas, on the border between Tibet and India and Nepal, as the "Roof of the World" (which would be Culmen Mundi in Latin), since Mt. Everest, at 29,035 ft., is the highest mountain in the world. There was no easy route across the Himalayas, however, so the term actually originates in Persian (Bam-e Donya) to refer to those ranges that were more familiar to travellers. The Pamirs are very nearly the center from which the other ranges radiate, and thus the focus of overland travel to and from India:

- The highest peak in the Pamirs, now in Tajikistan is Imeni Ismail Sumani Peak, formerly Communism Peak, formerly Stalin Peak (1933-1962), originally Garmo Peak, 24,590 ft. This was the highest point in the old Soviet Union.
- The highest peak in the Karakorum Range, on the border between Pakistani Kashmir and China, is K2

(Mount Godwin Austen), 28,250 ft. (second highest mountain in the world).

- The highest peak in the Hindu Kush, in Pakistan, is Tirich Mir, 25,230 ft.
- The highest peak in the Kunlun Range, in China, is Ulugh Muztagh (Mu-tzu-t'a-Ko), 25,340 ft.
- The highest peak in the Tien Shan (Tian Shan in Pinyin), on the border between China and Kyrgyzstan, is Pobedy Peak (Jengish Chokusu), 24,406 ft.

The Silk Road route from the Middle East to China mainly went through the Tarim Basin, either on the north side, below the Tian Shan, or the south side, above the Kunlun. The center of the Basin, the Taklimakan Desert, is waterless, uninhabited, and, really, uninhabitable. There was also a route north of the Tian Shan. That was a bit further, but the Dzungaria Basin, which I've heard described as a "sage brush and jack rabbit desert," is not as lifeless as the Taklimakan. Today, the through rail line and the larger cities (like Ürümqi) are on the north side on the Tian Shan.

All this is now in the Chinese territory of Sinkiang [Xinjiang]. The passes trough the Pamirs are so high, that familiar pack animals like horses and mules simply die. Only yaks are adapted to the altitude. But then yaks can't live at lower altitudes. Every caravan (from Perisan karvan), consequently, must begin down below with one kind of animal and then change over to yaks, and then back again once over the mountains. Perhaps the most famous Silk Road site, the caves full of Buddhist art and manuscripts at Dunhuang, is just north of the Nanshan ranges, still in the Chinese province of Kansu [Gansu].

Following Central Asia, the Andes have the highest peaks, culminating in Aconcagua at 22,834 ft. North America comes next, with Mt. McKinley at 20,320 ft. Then Africa, with Mt. Kilimanjaro at 19,340 ft. The Culmen Europae, the highest range in Europe, are the Caucasus mountains, whose highest peak is Mt. Elbrus (18,510 ft.). This is far from the population, historical, and cultural center of Europe. The Culmen Franciae is in the Alps, whose highest peak is Mt. Blanc at 15,771 ft.

This lay in the historic Kingdom of Burgundy but is now on the border between France and Italy.

Just to round things off, the highest peak in Antarctica is the Vinson Massif, at 16,864 ft. The highest peak in the remaining continent, Australia, is Mt. Kosciusko at only 7,310 ft. This is beat on two nearby islands, Mt. Jaya on New Guinea, at 16,500 ft., and Mt. Cook in New Zealand, at 12,349 ft. However, the highest peak in Polynesia is Mauna Kea on the Island of Hawai'i, at 13,796 ft. Measured from base to summit, Mauna Kea is itself actually the tallest mountain in the world. Mt. Everest is at the edge of the 10,000 foot Tibetan Plateau, meaning that it only rises about 19,000 feet. Mauna Kea, however, rises directly from the sea floor, which is at least 15,000 feet (three miles) down. Base to summit, Mauna Kea is at least 5.6 miles, or 29,568 ft. high (or perhaps as much as 33,000 feet, depending on the reference depth — sea level certainly makes this kind of thing easier).

Philosophy of History

Buddhist philosophy is the branch of Eastern philosophy based on the teachings of Gautama Buddha, a.k.a. Siddhartha Gautama (c. 563 BCE-c. 483 BCE). Buddhist philosophy deals extensively with problems in metaphysics, phenomenology, ethics, and epistemology.

From its inception, Buddhism has had a strong philosophical component. Buddhism is founded on the rejection of certain orthodox philosophical concepts, in which the Buddha had been instructed by various teachers. Buddhism rejects atheism, theism, monism, and dualism alike. The Buddha criticized all concepts of metaphysical being and non-being, and this critique is inextricable from the founding of Buddhism.

Particular points of Buddhist philosophizing have often been the subject of disputes between different schools of Buddhism. Metaphysical questions such as "Is there a god?" and "Does the soul (Atman) really exist?" have divided the Buddha's followers even during his own lifetime, and epistemological debates over the proper modes of evidence have always been lively in Buddhism.

Readers should note that theory for its own sake is not valued in Buddhism, but theory pursued in the interest of enlightenment for oneself or others is fully consistent with Buddhist values and ethics.

Buddhism as Philosophy

Some have asserted that Buddhism as a whole is a philosophy rather than a religion. Proponents of such a view may argue that (a) Buddhism is non-theistic (*i.e.*, it has no special use for the existence or nonexistence of a god or gods) or atheistic and (b) religions necessarily involve some form of theism. Others might contest either part of such an argument. Other arguments for Buddhism "as" philosophy may claim that Buddhism does not have doctrines in the same sense as other religions; the Buddha himself taught that a person should accept a teaching only if one's own experience verifies it and it is praised by the wise.

Arguments against Buddhism as a philosophy might call attention to the way Buddhism's pervasive inclusion of supernatural entities (not "gods" in the sense of Western monotheism, of course), to what most scholars identify as worship practices (ceremonial reverence of saints, etc.), to Buddhism's thoroughly developed hierarchies of clergy (not usually characteristic of a "philosophy"), and its overall religious organization.

A third perspective might take the position that Buddhism can be practiced either as a religion or as a philosophy. A similar distinction is often made with reference to Taoism.

Lama Anagorika Govinda expressed it as follows in the book 'A Living Buddhism for the West':

> *"Thus we could say that the Buddha's Dharma is, as experience and as a way to practical realisation, a religion; as the intellectual formulation of this experience, a philosophy; and as a result of self-observation and analysis, a psychology. Whoever treads this path acquires a norm of behaviour that is not dictated from without, but is the result of an inner process of*

maturation and that we-regarding it from without-can call morality."

It should also be noted that in the South and East Asian cultures in which Buddhism achieved most of its development, the distinction between philosophy and religion is somewhat unclear and possibly quite spurious, so this may be a semantic problem arising in the West alone.

Epistemology: Decisive in distinguishing Buddhism from what is commonly called Hinduism is the issue of epistemological justification. The schools of Indian logic recognize a certain set of valid justifications for knowledge, while Buddhism recognizes a smaller set. Both accept perception and inference, for example, but for the orthodox schools (of Hinduism), the received textual tradition (*e.g.*, the Vedas) is in itself an epistemological category equal to perception and inference (although this is not necessarily true for some of the non-orthodox schools, like Vedanta). Thus, in the orthodox schools, if a claim was made that could not be substantiated by appeal to the textual canon, it would be viewed as ridiculous as a claim that the sky was green.

Buddhism, on the other hand, rejected an inflexible reverence of accepted doctrine. As the Buddha said:

Do not accept anything by mere tradition... Do not accept anything just because it accords with your scriptures... Do not accept anything merely because it agrees with your pre-conceived notions... But when you know for yourselves—these things are moral, these things are blameless, these things are praised by the wise, these things, when performed and undertaken, conduce to well-being and happiness—then do you live acting accordingly.

Metaphysics and Phenomenology: In earliest Buddhism and today still in Theravda and the Madhyamaka, any metaphysical essence or being underlying the play of phenomenal experience is rejected. No "soul" or permanent self was recognized, and the perception of a continuous identity was held to be an illusion.

Any feeling whatsoever, any perception whatsoever, any mental processes whatsoever, any consciousness whatsoever—past, future, or present; internal or external; blatant or subtle, common or sublime, far or near; every consciousness—is to be seen as it actually is with right discernment as "This is not mine. This is not my self. This is not what I am." From within the context of the Madhyamaka, we find Candrakirti:

"Self is an essence of things that does not depend on others; it is an intrinsic nature. The non-existence of that is selflessness".

This anti-essentialist teaching, known as anatta, brought up many questions. If there is no atman or Brahman underlying the objects and events of the universe, how could they be explained? What gave them their existence? And if there was no "self", who makes the decisions we think we make, and what gets reincarnated? Early Buddhist philosophers and exegetes created a pluralist metaphysical and phenomenological system in which all experiences of people, things, and events, can be broken down into smaller and smaller perceptual or perceptual-ontological units called dharmas. These dharmas (roughly synonymous with "phenomena") were interpreted differently by different schools: some held they were real, some held only some were real, some held all were illusory, some held they were empty, some held they were intrinsically associated with suffering, etc.

Other debates in metaphysics and phenomenology include the issue of the Pudgala, or "person", which was inserted by the Pudgalavada school to replace the atman as that which transmigrates and that which carries the burden of karma from one life to another. Other schools made unsurprising objection to this. There were further sub-debates regarding whether the pudgala was real or illusory or something in between. The Yogacara school, somewhat later, would later elevate the mind to act as a substitute for Brahman, much as the Pudgala replaces the atman.

In many or all of these debates, some would point out the irony of pursuing questions which the Buddha was often prone to refuse to answer, on the grounds that they were non-conducive to enlightenment. For more detailed information, see Schools of Buddhism and the individual schools themselves.

Dependent Origination: The original positive Buddhist contribution to the field of metaphysics is pratityasamutpda, which arises from the Buddhist critique of Indian theories of causality. It states that events are not predetermined, nor are they random, and it rejects notions of direct causation owing to the need for such theories in the Indian context to be undergirded by a substantialist metaphysics. Instead, it posits the arising of events under certain conditions which are inextricable, such that the units in question at no time have independent existence.

This being, that becomes.
From the arising of this, that arises.
This not being, that does not become.
From the ceasing of this, that ceases. — Samyutta Nikaya

Pratitya-samutpada goes on to posit that certain specific events, concepts, or realities are always dependent on other specific things. Craving, for example, is always dependent on, and caused by, emotion. Emotion is always dependent on contact with our surroundings. This chain of causation purports to show that the cessation of decay, death, and sorrow is indirectly dependent on the cessation of craving, and ultimately dependent on an all-encompassing stillness.

Nagrjuna, one of the most influential Buddhist philosophers, asserted a direct connection between, even identity of, dependent origination, anatta and sunyat. He pointed out that implicit in the early Buddhist concept of dependent origination is the lack of any substantial being (anatta) underlying the participants in origination, so that they have no independent existence, a state identified as emptiness (sunyat), or emptiness of a nature or essence (sva-bhva). This element of Nagrjuna's thought is relatively uncontroversial, but it opens the way for his identification of samsara and nirvana, which was revolutionary.

Interpenetration: This doctrine comes from the Avatamsaka Sutra and its associated schools. It holds that all phenomena are intimately connected. Two images are used to convey this idea. The first is known as Indra's net. The net is set with jewels which have the extraordinary property that they reflect all of the other jewels. The second image is that of the world text. This image portrays the world as consisting of an enormous text which is as large as the universe itself. The 'words' of the text are composed of the phenomena that make up the world. However, every atom of the world contains the whole text within it. It is the work of a Buddha to let out the text so that beings can be liberated from suffering.

This idea was enormously influential on the Japanese monk kkkai in founding the Shingon school of Buddhism.

Ethics

Although there are many ethical tenets in Buddhism that differ depending on whether one is a monk or a layman, and depending on individual schools, the Buddhist system of ethics can always be summed up in the Eightfold Path. And this, monks, is the noble truth of the way of practice leading to the cessation of suffering — precisely this Noble Eightfold Path — right view, right resolve, right speech, right action, right livelihood, right effort, right mindfulness, right concentration.

The purpose of living an ethical life is to escape the suffering inherent in (unenlightened) worldly life. Although early Buddhism (Hinayana) is contrasted with later Buddhism (Mahayana) in that the latter emphasizes striving for the enlightenment of all (apparent) beings rather than simply oneself, in neither case can the motivation for ethical living be called 'selfish', because Buddhist doctrine holds the notion of a 'self' to be illusory. Buddhist teachings claim that there is no real difference between ourselves and others; therefore one should attempt to increase the happiness of all living things as eagerly as one's own. This is why many Buddhists choose to be vegetarians.

Early Development

The philosophical outlook of Earliest Buddhism was primarily negative, in the sense that it focused on what doctrines to reject more than on what doctrines to accept. This dimension has been preserved by the Madhyamaka school. It includes critical rejections of all views, which is a form of philosophy, but it is reluctant to posit its own. Only knowledge that is useful in achieving enlightenment is valued.

The cycle of philosophical upheavals that in part drove the diversification of Buddhism into its many schools and sects only began once Buddhists began attempting to make explicit the implicit philosophy of the Buddha and the early Suttas. After the death of the Buddha, attempts were made to gather his teachings and transmit them in a commonly agreed form, first orally, then also in writing (The Tripitaka). In addition to collecting the Buddha's speeches and rules for monastic life (Vinaya), monks soon undertook to condense what they considered the essential elements of Buddhist doctrine into lists of categories, provided with extensive commentary. This process took shape from about the 2nd century BCE to probably the 2nd century CE.

Early Buddhism: Before the royal sponsorship of Ashoka the Great in the 3rd century BCE, Buddhism seems to have remained a relatively minor phenomenon, and the historicity of its formative events is poorly established. Two formative councils are supposed to have taken place, although our knowledge of them is based on much later accounts. The councils tend to explain the formalization of the Buddhist doctrine, and the various subsequent schisms inside the Buddhist movement.

Buddhist Council: The first Buddhist council was held soon after the death of the Buddha under the patronage of king Ajatasatru of the Magadha empire, and presided by a monk named Mahakasyapa, at Rajagriha (today's Rajgir). The objective of the council was to record the Buddha's sayings (sutra) and codify monastic rules (vinaya): Ananda, one of the Buddha's main disciples and his cousin, was called upon to recite the discourses of the Buddha, and Upali, another disciple, recited the rules of the vinaya.

These became the basis of the Pali Canon, which has been the orthodox text of reference throughout the history of Buddhism.

The second Buddhist council was convened by King Kalasoka and held at Vaishali, following conflicts between the traditional schools of Buddhism and a more liberal interpretational movement called the Mahasanghikas. The traditional schools considered the Buddha as a human being who reached enlightenment, which could be most easily attained by monks following the monastic rules and practicing the teaching for the sake of overcoming suffering and attaining Arahantship.

The secessionist Mahasangikas, however, tended to consider this approach too individualistic and selfish. They considered the objective of becoming an arhat insufficient, and instead proposed that the only true goal was to reach full buddhahood, in a sense opening the way to future Mahyna thought. They became proponents of more relaxed monastic rules, which could appeal to a large majority of monastic and lay people (hence their name the "great" or "majority" assembly).

The council ended with the rejection of the Mahasanghikas. They left the council and maintained themselves for several centuries in northwestern India and Central Asia according to Kharoshti inscriptions found near the Oxus and dated c. 1st century CE.

Ashokan Proselytism

Fragment of the 6th Pillar Edict of Ashoka (238 BCE), in Brahmi, sandstone. British Museum.

The Mauryan emperor Ashoka the Great (273–232 BCE) converted to Buddhism after his bloody conquest of the territory of Kalinga (today's Orissa) in eastern India. Regretful of the horrors brought by the conflict, the king decided to renounce violence, and propagate the faith by building stupas and pillars urging for the respect of all animal life, and enjoining people to follow the Dharma. He also built roads, hospitals, resthouses, universities and irrigation systems around the country. He treated his subjects as equals regardless of their religion, politics or caste.

This period marks the first spread of Buddhism beyond India. According to the plates and pillars left by king Ashoka (the Edicts of Ashoka), emissaries were sent to various countries in order to spread Buddhism, as far as the Greek kingdoms in the West, in particular the neighboring Greco-Bactrian Kingdom, and possibly even farther to the Mediterranean according to the stone inscriptions left by Ashoka.

King Ashoka convened the third Buddhist council around 250 BCE at Pataliputra (today's Patna). It was held by the monk Moggaliputta. The objective of the council was to reconcile the different schools of Buddhism, to purify the Buddhist movement, particularly from opportunistic factions which had been attracted by the royal patronage, and to organize the dispatch of Buddhist missionaries throughout the known world.

The Pali canon (Tipitaka, or Tripitaka in Sanskrit, literally the "Three Baskets"), which comprises the texts of reference of traditional Buddhism and is considered to be directly transmitted from the Buddha, was formalized at that time. It consists of the doctrine (the Sutra Pitaka), the monastic discipline (Vinaya Pitaka) and an additional new body of subtle philosophy (the Abhidharma Pitaka).

The efforts of Ashoka to purify the Buddhist faith also had the effect of segregating against other emerging movements. In particular, after 250 BCE, the Sarvastivadin (who had been rejected by the 3rd council, according to the Theravada tradition) and the Dharmaguptaka schools became quite influential in northwestern India and Central Asia, up to the time of the Kushan Empire in the first centuries of the common era. The Dharmaguptakas were characterized by a belief that Buddha was separate, and above, the rest of the Buddhist community. The Sarvastivadin believed that past, present, and future are all simultaneous.

Hellenistic World: Some of the Edicts of Ashoka inscriptions describe the efforts made by Ashoka to propagate the Buddhist faith throughout the Hellenistic world, which at that time formed an uninterrupted continuum from the borders of India to Greece. The Edicts indicate a clear understanding of the political organization in Hellenistic territories: the names and location of the main Greek monarchs of the time are identified, and they are claimed as recipients of Buddhist proselytism: Antiochus II Theos of the Seleucid Kingdom (261–246 BCE), Ptolemy II Philadelphos of Egypt (285–247 BCE), Antigonus Gonatas of Macedonia (276–239 BCE), Magas of Cyrene (288–258 BCE), and Alexander II of Epirus (272–255 BCE).

Buddhist Proselytism at the Time of King Ashoka

"The conquest by Dharma has been won here, on the borders, and even six hundred yojanas (5,400-9,600 km) away, where the Greek king Antiochos rules, beyond there where the four kings named Ptolemy, Antigonos, Magas and Alexander rule, likewise in the south among the Cholas, the Pandyas, and as far as Tamraparni (Sri Lanka)." (Edicts of Ashoka, 13th Rock Edict, S. Dhammika).

Furthermore, according to Pali sources, some of Ashoka's emissaries were Greek Buddhist monks, indicating close religious exchanges between the two cultures:

"When the thera (elder) Moggaliputta, the illuminator of the religion of the Conqueror (Ashoka), had brought the (third) council to an end (...) he sent forth theras, one here and one there: (...) and to Aparantaka (the "Western countries" corresponding to Gujarat and Sindh) he sent the Greek (Yona) named Dhammarakkhita". (Mahavamsa XII).

Bilingual inscription (Greek and Aramaic) by king Ashoka, from Kandahar, Kabul Museum.

It is not clear how much these interactions may have been influential, but some authors have commented that some level of syncretism between Hellenist thought and Buddhism may have started in Hellenic lands at that time. They have pointed to the presence of Buddhist communities in the Hellenistic world around that period, in particular in Alexandria (mentioned by Clement of Alexandria), and to the pre-Christian monastic order of the Therapeutae (possibly a deformation of the Pali word "Theravada"), who may have "almost entirely drawn (its) inspiration from the teaching and practices of Buddhist asceticism" (Robert Lissen).

Buddhist gravestones from the Ptolemaic period have also been found in Alexandria, decorated with depictions of the Dharma wheel (Tarn, "The Greeks in Bactria and India"). Commenting on the presence of Buddhists in Alexandria, some scholars have even pointed out that "It was later in this very place that some of the most active centers of Christianity were established" (Robert Linssen "Zen living").

In the 2nd century CE, the Christian dogmatist Clement of Alexandria recognized Bactrian Buddhists (Sramanas) and Indian Gymnosophists for their influence on Greek thought:

> *"Thus philosophy, a thing of the highest utility, flourished in antiquity among the barbarians, shedding its light over the nations. And afterwards it came to Greece. First in its ranks were the prophets of the Egyptians; and the Chaldeans among the Assyrians; and the Druids among the Gauls; and the Sramanas among the Bactrians; and the philosophers of the Celts; and the Magi of the Persians, who foretold the Saviour's birth, and came into the land of Judaea guided by a star. The Indian gymnosophists are also in the number, and the other barbarian philosophers. And of these there are two classes, some of them called Sramanas and others Brahmins." Clement of Alexandria "The Stromata, or Miscellanies".*

Asian Expansion

In the areas east of the Indian subcontinent (today's Burma), Indian culture strongly influenced the Mons. The Mons are said to have been converted to Buddhism around 200 BCE under the proselytizing of the Indian king Ashoka, before the fission between Mahayana and Hinayana Buddhism. Early Mon Buddhist temples, such as Peikthano in central Burma, have been dated between the 1st and the 5thcentury CE.

Mons Wheel of the Law (Dharmacakra), art of Dvaravati, c.8th century.

The Buddhist art of the Mons was especially influenced by the Indian art of the Gupta and post-Gupta periods, and their mannerist style spread widely in South-East Asia following the expansion of the Mon kingdom between the 5th and 8th centuries. The Theravada faith expanded in the northern parts of Southeast Asia under Mon influence, until it was progressively displaced by Mahayana Buddhism from around the 6th century CE.

Sri Lanka was allegedly proselytized by Ashoka's son Mahinda and six companions during the 2nd century BCE. They converted the king Devanampiya Tissa and many of the nobility. This is when the Mahavihara monastery, a center of Sinhalese orthodoxy, was built. The Pali Canon was put in writing in Sri Lanka during the reign of king Vittagamani (r. 29–17 BCE), and the Theravada tradition flourished there, harbouring some great commentators such as Buddhaghosa (4th–5th century). Although Mahayana Buddhism gained some influence at that time, Theravada ultimately prevailed, and Sri Lanka turned out to be the last stronghold of Theravada Buddhism, from where it would expand again to South-East Asia from the 11th century.

There is also a legend, not directly validated by the edicts, that Ashoka sent a missionary to the north, through the Himalayas, to Khotan in the Tarim Basin, then the land of the Tocharians, speakers of an Indo-European language.

Rise of the Sunga (2nd–1st c.BCE)

Sunga Empire, Pusyamitra Sunga: The Sunga dynasty (185–73 BCE) was established in 185 BCE, about 50 years after Ashoka's death. After deposing King Brhadrata (last of the Mauryan rulers), military commander-in-chief Pusyamitra Sunga took the throne. An orthodox Brahmin, Sunga was allegedly hostile towards Buddhists and allegedly persecuted the Buddhist faith. He is recorded as having "destroyed monasteries and killed Monks" (Divyavadana, pp. 429–434): 84,000 Buddhist stupas which had been built by Ashoka were "destroyed" (R. Thaper), and 100 gold coins were offered for the head of each Buddhist monk (Indian Historical Quarterly Vol. XXII, p. 81 ff cited in Hars.407). A large number of Buddhist monasteries (viharas) were said to have been converted to Hindu temples, in such places as Nalanda, Bodhgaya, Sarnath, or Mathura. During the period, Buddhist monks deserted the Ganges valley, following either the Northern road (Uttarapatha) or the Southern road (Daksinapatha). Conversely, Buddhist artistic creation stopped in the old Magadha area, to reposition itself either in Northwest area of Gandhara and Mathura, or in the Southeast around Amaravati. Some artistic activity also occured in central India, as in Bharhut, to which the Sungas may or may not have contributed.

Silver Drachm of Menander-I

In the areas west of the Indian subcontinent, neighboring Greek kingdoms had been in place in Bactria (today's northern Afghanistan) since the time of the conquests of Alexander the Great around 326 BCE: first the Seleucids from around 323 BCE, then the Greco-Bactrian kingdom from around 250 BCE. A Greco-Buddhist statue, one of the first representations of the Buddha, 1st-2nd century CE, Gandhara.

The Greco-Bactrian king Demetrius I invaded India in 180 BCE as far as Pataliputra, establishing an Indo-Greek kingdom that was to last in various part of northern India until the end of the 1st century BCE. Buddhism flourished under the Indo-Greek kings, and it has been suggested that their invasion of India was intended to show their support for the Mauryan empire, and to protect the Buddhist faith from the alleged religious persecutions of the Sungas (185–73 BCE).

One of the most famous Indo-Greek kings is Menander (reigned c. 160–135 BCE). He apparently converted to Buddhism and is presented in the Mahayana tradition as one of the great benefactors of the faith, on a par with king Ashoka or the later Kushan king Kanishka. Menander's coins bear the mention "Saviour king" in Greek, and sometimes designs of the eight-spoked wheel. Direct cultural exchange is also suggested by the dialogue of the Milinda Panha between Menander and the monk Nagasena around 160 BCE. Upon his death, the honour of sharing his remains was claimed by the cities under his rule, and they were enshrined in stupas, in a parallel with the historic Buddha (Plutarch, Praec. reip. ger. 28, 6). Several of Menander's Indo-Greek successors inscribed the mention

"Follower of the Dharma" in the Kharoshthi script on their coins, and depicted themselves or their divinities forming the vitarka mudra.

A coin of Menander I with an eight-spoked wheel and a palm of victory on the reverse (British Museum).

The interaction between Greek and Buddhist cultures may have had some influence on the evolution of Mahayana, as the faith developed its sophisticated philosophical approach and a man-god treatment of the Buddha somewhat reminiscent of Hellenic gods. It is also around that time that the first anthropomorphic representations of the Buddha are found, often in realistic Greco-Buddhist style: "One might regard the classical influence as including the general idea of representing a man-god in this purely human form, which was of course well familiar in the West, and it is very likely that the example of westerner's treatment of their gods was indeed an important factor in the innovation" (Boardman, "The Diffusion of Classical Art in Antiquity").

Greco-Buddhism, Buddhism and the Roman World

Rise of Mahayana (1st c.BCE–2nd c.CE): The rise of Mahayana Buddhism from the 1st century BCE was accompanied by complex political changes in northwestern India. The Indo-Greek kingdoms were gradually overwhelmed, and their culture assimilated by the Indo-Scythians, and then the Yuezhi, who founded the Kushan Empire from around 12 BCE.

Coin of the Kushan emperor Kanishka, with the Buddha on the reverse, and his name "BODDO" in Greek script, minted circa 120 CE.

The Kushans were supportive of Buddhism, and a fourth Buddhist council was convened by the Kushan emperor Kanishka, around 100 CE at Jalandhar or in Kashmir, and is usually associated with the formal rise of Mahayana Buddhism and its secession from Theravada Buddhism. Theravada Buddhism does not recognize the authenticity of this council, and it is sometimes called the "council of heretical monks".

It is said that Kanishka gathered 500 bhikkhus in Kashmir, headed by Vasumitra, to edit the Tripitaka and make references and remarks. Allegedly, during the council there were all together three hundred thousand verses and over nine million statements compiled, and it took twelve years to complete. This council did not rely on the original Pali canon (the Tipitaka). Instead, a set of new scriptures was approved, as well as fundamental principles of Mahayana doctrine. The new scriptures, usually in the Gandhari vernacular and the Kharosthi script, were rewritten in the classical language of Sanskrit, to many scholars a turning point in the propagation of Buddhist thought.

The new form of Buddhism was characterized by an almost God-like treatment of the Buddha, by the idea that all beings have a Buddha-nature and should aspire to Buddhahood, and by a syncretism due to the various cultural influences within northwestern India and the Kushan Empire.

Mahayana Expansion (1st c. CE–10th c. CE): Expansion of Mahayana Buddhism between the 1st–10th century CE.

From that point on, and in the space of a few centuries, Mahayana was to flourish and spread in the East from India to South-East Asia, and towards the north to Central Asia, China, Korea, and finally to Japan in 538 CE.

India

After the end of the Kushans, Buddhism flourished in India during the dynasty of the Guptas (4th-6th century). Mahayana centers of learning were established, especially at Nalanda in north-eastern India, which was to become the largest and most influential Buddhist university for many centuries, with famous teachers such as Nagarjuna. The Gupta style of Buddhist art became very influential from South-East Asia to China as the faith was spreading there.

Buddha and Bodhisattvas, 11th Century, Pala Empire: Indian Buddhism had weakened in the 6th century following the White Hun invasions and Mihirkulas persecution.

Xuanzang reports in his travels across India during the 7th century of Buddhism being popular in Andhra, Dhanyakataka, and Dravida which today rougly correspond to the modern day

Indian states of Andhra Pradesh and Tamil Nadu. While reporting many deserted stupas in the area around modern day Nepal and the persecution of buddhists by Ssanka in the Kingdom of Gouda. (In modern day West Bengal.) Xuanzang compliments the patronage of Harshavardana during the same period.

After Harshavardanas kingdom, the rise of many small kingdoms that lead to the rise of the Rajputs across the gangetic plains and marked the end of Buddhist ruling clans along with a sharp decline in royal patronage until a revival under the Pala Empire in the Bengal region. Here Mahayana Buddhism flourished and spread to Bhutan and Sikkim between the 8th and the 12th century before the Palas collapsed under the assault of the Hindu Sena dynasty. The Palas created many temples and a distinctive school of Buddhist art. Xuanzang noted in his travels that in various regions Buddhism was giving way to Jainism and Hinduism.

By the 10th century Buddhism had experienced a sharp decline beyond the Pala realms in Bengal under a resurgent Hinduism and the incorporation in Vaishnavite Hinduism of Buddha as the 9th incarnation of Vishnu. A milestone in the decline of Indian Buddhism in the North occurred in 1193 when Turkic Islamic raiders under Muhammad Khilji burnt Nalanda. By the end of the 12th century, following the Islamic conquest of the Buddhist strongholds in Bihar, and the loss of political support coupled with social and caste pressures, the practice of Buddhism retreated to the Himalayan foothills in the North and Sri Lanka in the south. Additionally, the influence of Buddhism also waned due to Hinduism's revival movements such as Advaita, the rise of the bhakti movement and the missionary work of Sufis.

Central Asia

Central Asia had been influenced by Buddhism probably almost since the time of the Buddha. According to a legend preserved in Pali, the language of the Theravada canon, two merchant brothers from Bactria, named Tapassu and Bhallika, visited the Buddha and became his disciples. They then returned to Bactria and built temples to the Buddha (Foltz). Central Asia long played the role of a meeting place between China, India and Persia.

During the 2nd century BCE, the expansion of the Former Han to the west brought them into contact with the Hellenistic civilizations of Asia, especially the Greco-Bactrian Kingdoms. Thereafter, the expansion of Buddhism to the north led to the formation of Buddhist communities and even Buddhist kingdoms in the oases of Central Asia. Some Silk Road cities consisted almost entirely of Buddhist stupas and monasteries, and it seems that one of their main objectives was to welcome and service travelers between east and west.

The Hinayana traditions first spread among the Turkic tribes before combining with the Mahayana forms during the 2nd and 3rd centuries BCE to cover modern-day Pakistan, Kashmir, Afghanistan, eastern and coastal Iran, Uzbekistan, Turkmenistan and Tajikistan. These were the ancient states of Gandhara, Bactria, Parthia and Sogdia from where it spread to China. Among the first of these Turkic tribes to adopt Buddhism was the Turki-Shahi who adopted Buddhism as early as the 3rd century BCE. It was not, however, the exclusive faith of this region. There were also Zoroastrians, Hindus, Nestorian Christians, Jews, Manichaeans, and followers of shamanism, Tengrism, and other indigenous, nonorganized systems of belief.

Various Nikaya schools persisted in Central Asia and China until around the 7th century CE. Mahayana started to become dominant during the period, but since the faith had not developed a Nikaya approach, Sarvastivadin and Dharmaguptakas remained the Vinayas of choice in Central Asian monasteries. Various Buddhism kingdoms rose and prospered in both the Central Asian region and downwards into the Indian sub-continent such as Kushan Empire prior to the White Hun invasion in the 5th century where under the King Mihirkula they were heavily persecuted.

Buddhism in Central Asia started to decline with the expansion of Islam and the destruction of many stupas in war from the 7th century. The Muslims accorded them the status of dhimmis as "people of the Book", such as Christianity or Judaism and Al-Biruni wrote of Buddha as prophet "burxan".

Buddhism saw a surge during the reign of Mongols following the invasion of Genghis Khan and the establishment of the II Khanate and the Chagatai Khanate who brought their Buddhist influence with them during the 13th century, however within a 100 years the Mongols would convert to Islam and spread Islam across all the regions across central asia.

Mental Development (Samadhi)

The "mental development" subdivision of the Noble Eightfold Path is constituted by those elements that deal with how a Buddhist practitioner can best go about shaping his or her outlook towards the world.

Right Effort: Right effort (samyag-vyyma • samm-vyma) can also be translated as "right endeavor", and involves the Buddhist practitioner's continuous effort to, essentially, keep his or her mind free of thoughts that might impair his or her ability to realize or put into practice the other elements of the Noble Eightfold Path; for example, wishing ill towards another living being would contradict the injunction—contained in the "Right thought" element—to have good will towards others, and the "Right effort" element refers to the process of attempting to root out such an ill wish and replace it with a good wish. The Buddhist monk Ajahn Chah, of the Thai forest tradition of Theravada Buddhism, described right effort as follows:

Proper effort is not the effort to make something particular happen. It is the effort to be aware and awake in each moment, the effort to overcome laziness and defilement, the effort to make each activity of our day meditation.

By making right effort, a Buddhist practitioner is considered to be engaging in an effort that is wholesome in terms of karma; that is, in terms of that effort's ultimate consequences to the practitioner.

Right Mindfulness: Right mindfulness also translated as "right memory", together with right concentration, is concerned broadly with the practice of Buddhist meditation. Roughly speaking, "mindfulness" refers to the practice of keeping the mind alert to phenomena as they are affecting the body and mind. In the Magga-vibhanga Sutta, this aspect of the Noble Eightfold Path is explained as follows:

And what, monks, is right mindfulness?

(i) There is the case where a monk remains focused on (his/her) body in and of itself... ardent, aware, and mindful... having already put aside worldly desire and aversion.

(ii) (He/she) remains focused on feelings in and of themselves... ardent, aware, and mindful... having already put aside worldly desire and aversion.

(iii) (He/she) remains focused on the mind in and of itself... ardent, aware, and mindful... having already put aside worldly desire and aversion.

(iv) (He/she) remains focused on mental qualities in and of themselves... ardent, aware, and mindful... having already put aside worldly desire and aversion.

This, monks, is called right mindfulness.

Bhikkhu Bodhi, a monk of the Theravadin tradition, further glosses the concept of mindfulness as follows:

The mind is deliberately kept at the level of bare attention, a detached observation of what is happening within us and around us in the present moment. In the practice of right mindfulness the mind is trained to remain in the present, open, quiet, and alert, contemplating the present event. All judgments and interpretations have to be suspended, or if they occur, just registered and dropped.

Right Concentration: Right concentration together with right mindfulness, is concerned broadly with the practice of Buddhist meditation.

And what, monks, is right concentration?

(i) Quite withdrawn from sensuality, withdrawn from unwholesome states, a monk enters in the first jhna:

rapture and pleasure born from detachment, accompanied by movement of the mind onto the object and retention of the mind on the object.

(ii) With the stilling of directed thought and evaluation, he enters and remains in the second jhana: rapture and pleasure born of concentration; fixed single-pointed awareness free from movement of the mind onto the object and retention of the mind on the object; assurance.

(iii) With the fading of rapture, he remains in equanimity, mindful and fully aware, and physically sensitive of pleasure. He enters and remains in the third jhana which the Noble Ones declare to be "Equanimous and mindful, (he/she) has a pleasurable abiding."

(iv) With the abandoning of pleasure and pain...as with the earlier disappearance of elation and distress...he enters and remains in the fourth jhana: purity of equanimity and mindfulness, neither in pleasure nor in pain.

This, monks, is called right concentration.

The Ninth and Tenth Elements: In the Great Forty Sutra (Mahcattrisaka Sutta), which appears in the Pali Canon, the Buddha explains that cultivation of the Eightfold Path leads to the development of two further stages once enlightenment has been reached. These also fall under the category of panna and are Right Knowledge (sammnna) and Right Liberation (or Right Release; sammvimutti). Some consider Right Association as an implicit ninth aspect of the Path.

The Noble Eightfold Path and Cognitive Psychology: From the standpoint of modern cognitive psychology, the Noble Eightfold Path can be seen as rooted in what is called cognitive dissonance, which is the perception of incompatibility between two cognitions. In the essay "Buddhism Meets Western Science", Gay Watson explains this dissonance as it relates to Buddhist teaching:

Buddhism has always been concerned with feelings, emotions, sensations, and cognition. The Buddha points both to cognitive and emotional causes of suffering. The emotional cause is desire and its negative opposite, aversion. The cognitive cause is ignorance of the way things truly occur, or of three marks of existence: that all things are unsatisfactory, impermanent, and without essential self.

The Noble Eightfold Path is, from this psychological viewpoint, an attempt to resolve this dissonance by changing patterns of thought and behaviour. It is for this reason that the first element of the path is right understanding (samma-ditthi), which is how one's mind views the world. Under the wisdom (panna) subdivision of the Noble Eightfold Path, this worldview is intimately connected with the second element, right thought (samma-samkappa), which concerns the patterns of thought and intention that controls one's actions. These elements can be seen at work, for example, in the opening verses of the Dhammapada:

Preceded by perception are mental states,
For them is perception supreme,
From perception have they sprung.
If, with perception polluted, one speaks or acts,
Thence suffering follows
As a wheel the draught ox's foot.
Preceded by perception are mental states,
For them is perception supreme,
From perception have they sprung.
If, with tranquil perception, one speaks or acts,
Thence ease follows
As a shadow that never departs.

Thus, by willfully altering one's distorted worldview as well as the behaviors stemming from that worldview and bringing out "tranquil perception" in the place of "perception polluted", one is enabled to potentially escape from suffering and develop one's mind. Watson points this out from a psychological standpoint:

Research has shown that repeated action, learning, and memory can actually change the nervous system physically, altering both synaptic strength and connections. Such changes may be brought about by cultivated change in emotion and action; they will, in turn, change subsequent experience.

As such, Buddhism can essentially be seen as mind cultivation and rehabilitation.

Panchasila

Pancha or Pancasila (Pli) means the five precepts, which are the fundamental code of ethics, undertaken by lay followers of the Buddha Gautama.

The laity undertake to follow these training rules at the same time as they become Buddhists, taking refuge in the Triple Gem: In the Buddha (teacher), in the Dharma (teaching) and thirdly in the Sangha (community of monks and nuns). The Buddha taught the five precepts out of compassion, not out of any desire to control his followers, and so they are to be undertaken voluntarily rather than as "commandments" from a god. The following are rendered in English and then Pali.

1. I undertake the precept to refrain from destroying living beings. Pntipt veramani sikkhpadam samdiymi
2. I undertake the precept to refrain from stealing. Adinndn veramani sikkhpadam samdiymi
3. I undertake the precept to refrain from sexual misconduct (adultery, rape, etc). Kmesu micchcra veramani sikkhpadam samdiymi
4. I undertake the precept to refrain from false speech (lying). Musvda veramani sikkhpadam samdiymi
5 I undertake the precept to refrain from intoxicants which lead to heedlessness. Sur meraya majja pamdatthn veramani sikkhpadam samdiymi

Nirvana

Nirvana is, further, the abiding of a fully enlightened being (see terms Arhat and Buddha) in pure unobstructed awareness of reality as-it-is. It is accompanied by and synonymous with a state of spiritual awakening (Enlightenment). The Buddha describes the abiding in nirvana as a state of 'deathlessness' (Pali: amata or amaravati) and the "unconditioned" and as the highest spiritual attainment, the natural result that accrues to one who lives a life of virtuous conduct in accordance with Dharma.

There are many synonyms for Nirvana, as shown by the following passage:

> *World Honored One, the ground of fruition is bodhi, nirvana, true suchness, the Buddha-nature, the amala-consciousness, the empty treasury of the Thus Come One, the great, perfect mirror-wisdom. But although it is called by these seven names, it is pure and perfect, its substance is durable, like royal vajra, everlasting and indestructible. (Surangama Sutra IV 207)*

Nirvna (Pali nibbna) in sutra is "bhavanirodha nibbnam" (The cessation of becoming means Nirvna). Nirvna in sktra is never conceived of as a place, but the antinomy of samsra (see below) which itself is synonymous with ignorance (avidy, Pli avijj). "This said:

> *'the liberated mind/will (citta) which does not cling' means Nibbna" (Majjhima Nikaya 2-Att. 4.68).*

Nibbna is meant specifically as pertains gnosis that which ends the identity of the mind (citta) with empirical phenomena. Doctrinally Nibbna is said of the mind which no "longer is coming (bhava) and going (vibhava)", but which has attained a status in perpetuity, whereby "liberation (vimutta) can be said".

It carries further connotations of stilling, cooling, and peace; the realizing of nirvana is compared to the ending of avidy (ignorance) which perpetuates the will (citta/mind) from passing through samśra life after life, which causes (and is caused by) among other things craving, consciousness, birth, death, greed, hate, delusion, ignorance. Nirvna, then, is not a place nor a state, it is an absolute truth to be realized, and a person can do so without dying. When a person who has realized nirvna dies, his death is referred as his parinirvna, his fully passing away, as his life was his last link to the cycle of death and rebirth (samsara), and he will not be reborn again. Buddhism holds that the ultimate goal and end of samsaric existence (of ever "becoming" and "dying" and never truly being) is realization of Nirvana; what happens to a person after his parinirvana cannot be explained, as it is outside of all conceivable experience.

Undefinable Nature: Ergo the freed mind is equal to Nibbana in Buddhist doctrine. Elsewhere the Buddha calls Nirvana 'the unconditioned element' (*i.e.*, that which is not subject to causation). Nirvana in doctrinal citation the "subjugation of becoming" (bhavanirodha nibbanam) and is relavent to Ignorance (avijja) only, such that Samsara is the will/mind (citta) in ignorance, and Nirvana/Nibbana the will/mind devoid of said ignorance; it can only be experienced in dirrect Subjective gnosis. While some of the associated effects of Nirvana can be identified, a definition of Nirvana can only be approximated by what it is not. It is not the clinging existence with which man is understood to be afflicted. It is not any sort of becoming. It has no origin or end. It is not made or fabricated. It has no dualities, so that it cannot be described in words. It has no parts that may be distinguished one from another. It is not a subjective state of consciousness. It is not conditioned on or by anything else.

It should also be noted that the Buddha discouraged certain lines of speculation, including speculation into the state of an enlightened being after death, on the grounds that these were not useful for pursuing enlightenment; thus definitions of Nirvana might be said to be doctrinally unimportant.

In the Samyutta Nikaya, the Buddha describes Nibbna as:

> *"the far shore, the subtle, the very difficult to see, the unaging, the stable, the undisintegrating, the unmanifest, the unproliferated, the peaceful, the deathless, the sublime, the auspicious, the secure, the destruction of craving, the wonderful, the amazing, the unailing, the unailing state, the unafflicted, dispassion, purity, freedom, the unadhesive, the island, the shelter, the asylum, the refuge..."*

At the end of the Mahsatipammhana Sutta in Digha Nikya, the Buddha describes Success of Four Pammhana Meditations as: "One who is honest to himself and practice this four

Pammhana Meditations without a delay, he should be willing to achieve Arahat or Angami level, in seven days to seven years in time which would ultimately direct to Nirvana."

What Nirvana is Not?

Nirvana is the complete realization of the middle way (majjhena dha"mma"m deseti) and is neither the extremist views of nihilism (Pali: Ucchedavaada) nor eternalism (Pali: Sassatavaada) nor the monism of "oneness with brahman".

In the Samannaphala Sutta of the Digha Nikaya, Buddha clearly outlines the differences between his teachings and these schools which are considered pasanda or schools which lead to wrong views. Nirvana is not eternalism as Buddhism posits Anatta (not-self), so there is no personal immortality of a self.

Nirvana is not Nihilism: Which identifies the psycho-physical person (nama-rupa) with the body (rupa), rejecting human effort and the world hereafter (para loka). When the body is dead, it entails the total annihilation of the psycho-physical person, without the continuity of the consciousness for bearing moral retribution of his deeds done;

Similarly, Nirvana is not equivalent to the Vedic concept of Brahman, nor is the same as Heaven in many other religions:

> *The Buddha emphasized that the atman is like a mountain stream, which flows fast and is forever changing. There is no being (sat), there is only becoming (bhava) in it. The arising (uppaada), disappearance (vyaya) and changing of what exists (a~n~natatha) are the three signs of compounded things. The belief in a permanent self (atman) not only negates the activities of moral life but also falls in a form of grasping, a hindrance to spiritual liberation.*

In more precise terms, it is the non-Self (*i.e.*, the mundane body and unawakened mind, not the "Self") which is, according to the pronouncements of the Buddha in the Nikayas, linked to impermanence and change. Clutching onto this non-Self and the latter's suffering-generating mutations is what deprives sentient beings of the attainment of Nirvana.

Nirvana and Samsara: In Mahayana Buddhism, calling Nirvana the "opposite" of samsara or implying that it is apart from samsara is doctrinally problematic. According to early Mahayana Buddhism, they can be considered to be two aspects of the same perceived reality. By the time of Nagarjuna, there are teachings of the identity of Nirvana and samsara. However, even here it is assumed that the natural man suffers from at the very least a confusion regarding the nature of samsara.

The Theravda school makes the antithesis of samsara and Nibbana the starting point of the entire quest for deliverance. Even more, it treats this antithesis as determinative of the final goal, which is precisely the transcendence of samsara and the attainment of liberation in Nibbana. Where Theravada differs significantly from the Mahayana schools, which also start with the duality of samsara and Nirvana, is in not regarding this polarity as a mere preparatory lesson tailored for those with blunt faculties, to be eventually superseded by some higher realization of non-duality. From the standpoint of the Pali Suttas, even for the Buddha and the Arahants suffering and its cessation, samsara and Nibbana, remain distinct.

In the experience of some, Nirvana is a state which all six bases (Eye, Ear, Nose, Tongue, Body and Mind) cannot feel.

It is probably best to understand the relationship between Nirvana and samsara in terms of the Buddha while on earth. Buddha was both in Samsara while having attained to Nirvana so that he was seen by all, and simultaneously free from samsara.

Nirvana in Buddhist Commentaries: Sarvastivadin commentary, Abhidharma-mahavibhasa-sastra, gives the complete context of the possible meanings from it's sanskrit roots:

- Vana, implying the path of rebirth, + nir, meaning leaving off' or 'being away from the path of rebirth.'
- Vana, meaning 'stench', + nir, meaning 'freedom': 'freedom from the stench of distressing kamma.'
- Vana, meaning 'dense forests', + nir, meaning 'to get rid of' = 'to be permanently rid of the dense forest of the five aggregates (panca skandha), or the 'three roots

of greed, hate and delusion (lobha, dosa, moha)' or 'three characteristics of existence (impermanence, anitya; unsatisfactoriness, dukkha, soullessness, anatma).

- Vana, meaning 'weaving', + nir, meaning 'knot' = 'freedom from the knot of the distressful thread of kamma.'

Nirvana in the Mahparinirvana Sutra: The nature of Nirvana assumes a differently aspected Mahayana focus in what alleges to be the final of all Mahayana sutras, allegedly delivered by the Buddha on his last day of life on earth-the Mahaparinirvana Sutra or Nirvana Sutra. Here, as well as in a number of linked "tathagatagarbha" sutras, in which the Tathagatagarbha is equated with the Buddha's eternal Self or eternal nature, Nirvana is spoken of by the Mahayana Buddha in very "cataphatic", positive terms. Nirvana, or "Great Nirvana", is indicated to be the sphere or domain (vishaya) of the True Self. It is seen as the state which constitutes the attainment of what is "Eternal, the Self, Bliss, and the Pure". Mah-Nirvana ("Great Nirvana") thus becomes equivalent to the ineffable, unshakeable, blissful, all-pervading and deathless Selfhood of the Buddha himself-a mystery which no words can adequately reach and which, according to the Nirvana Sutra, can only be fully known by an Awakened Being-a perfect Buddha-directly.

Strikingly, the Buddha of the Mahaparinirvana Sutra gives the following definition of the attributes of Nirvana, which includes the ultimate reality of the Self (not to be confused with the "worldly ego" of the 5 skandhas):

"The attributes of Nirvana are eightfold. What are these eight? Cessation [nirodha], loveliness / wholesomeness [subha], Truth [satya], Reality [tattva], eternity [nitya], bliss [sukha], the Self [atman], and complete purity [parisuddhi]: that is Nirvana."

He further states: "Non-Self is Samsara [the cycle of rebirth]; the Self (atman) is Great Nirvana."

An important facet of Nirvana in general is that it is not something that comes about from a concatenation of causes, that springs into existence as a result of an act of creation or an agglomeration of causative factors: it was never created; it always was, is and will be.

But due to the moral and mental darkness of ordinary, samsarically benighted sentient beings, it remains hidden from unawakened perception. The Buddha of the Mahparinirvanasutra insists on its eternal nature and affirms its identity with the enduring, blissful Self, saying:

> *"It is not the case that the inherent nature of Nirvana did not primordially exist but now exists. If the inherent nature of Nirvana did not primordially exist but does now exist, then it would not be free from taints (sravas) nor would it be eternally (nitya) present in nature. Regardless of whether there are Buddhas or not, its intrinsic nature and attributes are eternally present... Because of the obscuring darkness of the mental afflictions (kle[as), beings do not see it. The Tathgata, endowed with omniscient awareness (sarvajn-jnana), lights the lamp of insight with his skill-in-means (upya-kau[alya) and causes Bodhisattvas to perceive the Eternal, Bliss, the Self, and the Pure of Nirvana."*

Vitally, according to these Mahayana teachings, any being who has reached Nirvana is not blotted out or extinguished: there is the extinction of the impermanent and suffering-prone "worldly self" or ego (comprised of the five changeful skandhas), but not of the immortal "supramundane" Self of the indwelling Buddha Principle [Buddha-dhatu]. Spiritual death for such a Nirvana-ed being becomes an utter impossibility.

The Buddha states in the "Mahayana Mahparinirvana Sutra" (Tibetan version): "Nirvana is deathless... Those who have passed into Nirvana are deathless. I say that anybody who is endowed with careful assiduity is not compounded and, even though they involve themselves in compounded things, they do not age, they do not die, they do not perish."

Quotations

- Gautama Buddha:
- "Nirvana is the highest happiness."

- "Where there is nothing; where naught is grasped, there is the Isle of No-Beyond. Nirvana do I call it — the utter extinction of aging and dying."
- "There is, monks, an unborn — unbecome — unmade— unfabricated. If there were not that unborn— unbecome— unmade — unfabricated, there would not be the case that emancipation from the born — become— made — fabricated would be discerned. But precisely because there is an unborn — unbecome — unmade — unfabricated, emancipation from the born — become— made — fabricated is discerned."
- This said: 'the liberated mind/will (citta) which does not cling' means Nibbana"
- "The subjugation of becoming means Nirvana'; this means the subjugation of the five aggregates means Nirvana."
- "Parinibbuto thitatto"-"Parinirvana is to be fixed in the Soul"
- Said immediately after the physical death of Gotama Buddha wherein his mind (citta) is= parinirvana=the essence of liberation:
- "No longer with (subsists by) in-breath nor out-breath, so is him (Gotama) who is steadfast in mind (citta), inherently quelled from all desires the mighty sage has passed beyond. With mind (citta) limitless (Brahman) he no longer bears sensations; illumined and unbound (nibbana), his mind (citta) is definitely (ahu) liberated."
- "The mind (citta) being so liberated and arisen from defilements, one is fixed in the Soul as liberation, one is quelled in fixation upon the Soul. Quelled in the Soul one is unshakable. So being unshakable, the very Soul is thoroughly unbound (parinirvana)."
- Sutta Nipata, tr. Rune Johansson:
- Like a flame that has been blown out by a strong wind goes to rest and cannot be defined, just so the sage who is freed from name and body goes to rest and cannot be defined.

For him who has gone to rest there is no measure by means of which one could describe him; that is not for him. When all (dharmas) have gone, all signs of recognition have also gone.

- Venerable Sariputta:
- The destruction of greed, hatred and delusion is Nirvana.

Three Jewels

The Three Jewels, also rendered as Three Treasures, Three Refuges or Triple Gem (Sanskrit: Triratna, also Ratna-traya, Pali: Tiratana, Tisarana are the three things that Buddhists give themselves to, and in return look toward for guidance, in the process known as taking refuge.

Taking the the Triple Gem is generally considered makes one officially a Buddhist. Thus, in many Buddhist communities, the following Pali chant, the Vandana Ti-sarana is often recited by both monks and lay people:

- Buddham Sharanam Gacchami I go for refuge in the Buddha
- Dhammam Sharanam Gacchami I go for refuge in the Dharma
- Sangham Sharanam Gacchami I go for refuge in the Sangha

Taking refuge in the Three Jewels is central to Buddhist lay and monastic ordination ceremonies, as originated by Gautama Buddha.

These three types of persons are found in the world: One with a mind like an open sore; one with a mind like a flash of lightning; one with a mind like a diamond.

- One who is irascible and very irritable, displaying anger, hatred and sulkiness; such a one is said to be a person with a mind like an open sore.
- One who understands the Four Noble Truths correctly is said to have a mind like a flash of lightning.
- One who has destroyed the mind-intoxicating defilements and realized the liberation of mind and the liberation by knowledge is said to have a mind like a diamond

With this we understand that to take refuge in the Buddha is to take refuge in the mind like a diamond, the hardest natural substance that can cut through all delusion.

The Three Gems when used in the process of taking refuge, become the Three Refuges.

The expression Three Gems are found in the earliest Buddhist literature of the Pali Canon, besides other works there is one sutta in the Sutta-nipata, called the Ratana-sutta which contains a series of verses on the Jewels in the Buddha, Dharma, and Sangha. In the Ratana-sutta, all the qualities of the Sangha mentioned are attributes of the Buddha's enlightened disciples.

The Three Refuges occur very frequently in the ancient Buddhist Texts, and here the Sangha is used more broadly to refer to either the Sangha of Bhikkhus, or the Sangha of Bhikkhunis.

"I go to Master Gotama for refuge and to the Dhamma, and to the Sangha of Bhikkhus."

Triratna Symbol: The compound Buddhist symbols: Shrivatsa within a triratana, over a Dharmacakra wheel, on the Tonana gate at Sanchi. 1st century BCE.

The Three Jewels are also symbolized by the triratna, composed of (from bottom to top):

- A lotus flower within a circle.
- A diamond rod, or vajra.
- A trident, or trisula, with three branches, representing the threefold jewels of Buddhism: Buddha, the Dharma and the Sangha.

On representations of the footprint of the Buddha, the Triratna is usually also surmounted by the Dharma wheel.

The Triratna ćan be found on frieze sculptures at Sanchi as the symbol crowning a flag standard (2nd century BCE), as a symbol of the Buddha installed on the Buddha's throne (2nd century BCE), as the crowning decorative symbol on the later gates at the stupa in Sanchi (2nd century CE), or, very often on the Buddha footprint (starting from the 1st century CE).

2nd century BCE coin of the Kunindas, incorporating on the reverse the Buddhist triratna symbol on top of a stupa.

The Triratna is also on the 1st century BCE coins of the Kingdom of Kuninda in northern Punjab, surmounting depictions of stupas, on some the coins of the Indo-Parthian king Gondophares, or the coins of some of the Kushan kings such as Vima Kadphises. The triratna can be further reinforced by being surmounted with three dharma wheels (one for each of the three jewels of Buddhism: the Buddha, the Dharma and the Sangha).

The triratna symbol is also called nandipada, or "bull's hoof", by Hindus.

Qualities of the Buddha, Dharma, and Sangha: The Triratna or "Three Jewels" symbol, on a Buddha footprint. 1st century CE, Gandhara. The qualities of the Buddha, Dharma, and Sangha are frequently repeated in the ancient texts. It is a Buddhist practise to reflect upon them.

- *The Buddha:* "The Blessed One is an Arahant, perfectly enlightened, accomplished in true knowledge and conduct, fortunate, knower of the world, unsurpassed leader of persons to be tamed, teacher of devas and humans, the Enlightened One, the Blessed One."
- *The Dharma:* "The Dhamma is well expounded by the Blessed One, directly visible, immidiate, inviting one to come and see, applicable, to be personally experienced by the wise."
- *The Sangha:* "The Sangha of the Blessed One's disciples is practising the good way, practising the straight way, practising the true way, practising the proper way; that is, the four pairs of persons, the eight types of individuals-This Sangha of the Blessed One's disciples is worthy of gifts, worthy of hospitality, worthy of offerings, worthy of reverential salutation, the unsurpassed field of merit for the world."

Three marks of existence:

After much meditation, the Buddha concluded that everything in the physical world (plus everything in the phenomenology of psychology) is marked by three characteristics, known as the three characteristics of existence, three signs of being or Dharma Seals. Together the three characteristics of existence are called ti-lakkhana, in Pali; or tri-laksana, in Sanskrit.

- Dukkha or unsatisfactoriness. Nothing found in the physical world or the psychological realm can bring lasting deep satisfaction.
- Anicca or impermanence. This refers not only to the fact that all conditioned things eventually cease to exist, but also that all conditioned things are in a constant state of flux. (A convenient way to visualize this would be to recall that the atoms constituting your body are constantly being replaced.)
- Anatta, impersonality, or non-Self. The human personality, "soul", or Self, is a conventional appellation applied to the assembly of physical and psychological components, each individually subject to constant flux; there is no central core (or essence); this is somewhat similar to a bundle theory of mind or soul.

There is often a fourth Dharma Seal mentioned:

- Nirvana is peace. Nirvana is the 'other shore' from Samsara.

By bringing the three (or four) seals into moment-to-moment experience through concentrated awareness, we are said to achieve Wisdom-the third of the three higher trainings-the way out of Samsara. In this way we can identify that, according to Sutra, the recipe (or formula) for leaving Samsara is achieved by a deep-rooted change to our Weltanschauung.

According to the Buddhist tradition, all phenomena (dharmas) are marked by three characteristics, sometimes referred to as the Dharma seals, that is dukkha (suffering), anicca (impermanence), and anatta (non-Self).

Dukkha

Whatever is impermanent is subject to change. Whatever is subject to change is subject to suffering.— The Buddha

Striving for what we desire, we may experience stress and suffering—dukkha. Getting what we desired, we may find delight and happiness. Soon after, the novelty may wear out and we may get bored with it. Boredom is a form of dissatisfaction (or suffering) and to escape from it, we divert ourselves from such boredom by indulging in a pursuit of new forms of pleasure. Sometimes not willing to relinquish objects that we are already disinterested in, we start to collect and amass possessions instead of sharing with others who may have better use in it than we do. Boredom is a result of change. Change of our interest in that object of desire that so captivated us in the first place.

If we do not get bored already, then change may instead occur in the object of desire. Silverware may becom tarnished, a new dress worn thin or a gadget gone obsolete. Or it may become broken, causing us to grieve. In some cases it may get lost or stolen. In some cases, we may worry about such losses even before they happen. Husbands and wives worry about losing their spouses even though their partners are faithful. Unfortunately, sometimes our very worry and fear drives us to act irrationally, resulting in distrust and breaking up of the very relationship that we cherished so much.

While we like changes such as becoming an adult when we are in our teens, we dislike the change called aging. While we strive for change to become rich, we fear the change of retrenchment. We are selective in our attitude towards the transient nature of our very existence. Unfortunately, this transient nature is unselective. We can try to fight it, just as many have tried since beginningless time, only to have our efforts washed away through the passages of time. As a result, we continually experience dissatisfaction or suffering due to the very impermanence of compounded phenomena.

Only in the realm of Nirvana—so Mahayana Buddhism insists—can true and lasting happiness be found. Nirvana is the opposite of the conditional, the transitory and the painful (dukkha), so it does not result in disappointment or deterioration of the state of bliss. Nirvana is the refuge from the otherwise universal tyranny of change and suffering. In other schools of

Buddhism, nirvana is not viewed as the goal, but merely as a projection from the state of samsara. According to these schools samsara (confusion) and nirvana (perfection) are two sides of the same coin that must be transcended through diligent practice of meditation.

Anicca

All compounded phenomena (things and experiences) are inconstant, unsteady, and impermanent. (Practically) everything is made up of parts, and is dependent on the right conditions for its existence. Everything is in constant flux, and so conditions and the thing itself is constantly changing. Things are constantly coming into being, and ceasing to be. Nothing lasts.

The important point here is that phenomena arise and cease according to (complex) conditions and not according to our whims and fancy. While we have limited ability to effect change to our possessions and surroundings, experience tells us that our feeble attempts are no guarantee that the results of our efforts will be to our likings. More often than not, the results fall short of our expectations.

In Mahayana Buddhism, a caveat is added: one should indeed always meditate on the impermanence and changefulness of compounded structures and phenomena, but one must guard against extending this to the realm of Nirvana, where impermanence holds no sway and eternity alone obtains. To see Nirvana or the Buddha (in his ultimate Dharmakaya nature) as impermanent would be to indulge in "perverted Dharma" and would be seriously to go astray, according to the Buddha's final Mahayana doctrines. Other schools of Buddhism, however, feel uneasy with such a teaching.

Anatta

In Indian philosophy, the concept of a Self is called atman (that is, "soul" or metaphysical Self), which refers to an unchanging, permanent essence conceived by virtue of existence. This concept and the related concept of Brahman, the Vedantic monistic ideal, which was regarded as an ultimate atman for all beings, were indispensable for mainstream Indian metaphysics, logic, and science; for all apparent things there had to be an underlying and persistent reality, akin to a Platonic form. The Buddha rejected all concepts of atman, emphasizing not permanence, but changeability. He taught that all concepts of a substantial personal Self were incorrect, and formed in the realm of ignorance.

However, in a number of major Mahayana sutras (*e.g.* the Mahaparinirvana Sutra, the Tathagatagarbha Sutra, the Srimala Sutra, among others), the Buddha is presented as clarifying this teaching by saying that, while the skandhas (constituents of the ordinary body and mind) are not the Self, there does truly exist an eternal, unchanging, blissful Buddha-essence in all sentient beings, which is the uncreated and deathless Buddha-nature ("Buddha-dhatu") or "True Self" of the Buddha himself. This immaculate Buddhic Self (atman) is in no way to be construed as a mundane, impermanent, suffering "ego", of which it is the diametrical opposite. On the other hand, this Buddha-essence or Buddha-nature is also often explained as the potential for achieving Buddhahood, rather than an existing phenomenon one can grasp onto as being me or Self.

It is the opposite of a personalised, samsaric "I" or "mine". The paradox is that as soon as the Buddhist practitioner tries to grasp at this inner Buddha potency and cling to it as though it were his or her ego writ large, it proves elusive. It does not "exist" in the time-space conditioned and finite mode in which mundane things are bodied forth. It is presented by the Buddha in the relevant sutras as ultimately inexplicable, primordially present Reality itself-the living potency for Buddhahood inside all beings. It is finally revealed (in the last of the Buddha's Mahayana sutras, the Nirvana Sutra) not as the circumscribed "non-Self", the clinging ego (which is indeed anatta/anatman), but as the ever-enduring, egoless Great Self or Dharmakaya of the Buddha.

The scriptural evidence of the Nikyas and gamas is ambivalent with regard to the Buddha's reported views on the existence or otherwise of a permanent Self (atman/atta). Though he is clearly reported to have criticized many of the heterodox concepts concerning an eternal personal Self and to have denied the existence of an eternal Self with regards to any of the constituent elements (skandha) of a being, he is nevertheless not reported to have explicitly denied the existence of a non-personal, permanent Self, contrary to the popular, orthodox view of the Buddha's teachings. Moreover, when the Buddha predicates "antman" (anatta) with regards to the constituents of a being, there is a grammatical ambivalence in the use of the term.

The most natural interpretation is that he is simply stating that "the constituents are not the Self" rather than "the constituents are devoid of Self". This ambivalence was to prove troublesome to Buddhists after the Buddha's passing. Some of the major schools of Buddhism that developed subsequently maintained the former interpretation, but other influential schools adopted the latter interpretation and took measures to establish their view as the orthodox Buddhist position.

One such proponent of this hard-line "non-Self" position was the monk Nagasena, who appears in the Questions of King Milinda, composed during the period of the Hellenistic Indo-Greek kingdom of the 2nd and 1st centuries BCE. In this text, Nagasena demonstrates the concept of absolute 'non-Self' by likening human beings to a chariot and challenges the Greek king "Milinda" (Menander) to find the essence of the chariot. Nagasena states that just as a chariot is made up of a number of things, none of which are the essence of the chariot in isolation, without the other pieces, similarly no one part of a person is a permanent entity; we can be broken up into five constituents — body, sensations, ideation, mental formations and consciousness— the consciousness being closest to the permanent idea of 'Self', but is ever-changing with each new thought according to this view point.

According to some thinkers both in the East and the West, the doctrine of "non-Self", may imply that Buddhism is a form of nihilism or something similar. However, as thinkers like Nagarjuna have clearly pointed out, Buddhism is not simply a rejection of the concept of existence (or of meaning, etc.) but of the hard and fast distinction between existence and non-existence, or rather between being and nothingness. Phenomena are not independent from causes and conditions, and do not exist as isolated thing as we perceive them to be. Philosophers such as N stress that the lack of a permanent, unchanging, substantial Self in beings and things does not mean that they do not experience growth and decay on the relative level. But on the ultimate level of analysis, one cannot distinguish an object from its causes and conditions, or even object and subject. (This is an idea appearing relatively recently in Western science.) Buddhism thus has much more in common with Western empiricism, pragmatism, anti-foundationalism, and even poststructuralism than with nihilism.

In the Nikyas, the Buddha and his disciples are commonly found to ask in question or declare "Is that which is impermanent, subject to change, subject to suffering fit to be considered thus: 'This I am, this is mine, this is my self'?" The question which the Buddha posts to his audience is whether compounded phenomena is fit to be cosidered as self, in which the audience agrees that it is unworthy to be considered so. And in relinquishing such an attachment to compounded phenomena, such a person gives up delight, desire and craving for compounded phenomena and is unbounded by its change. When completely free from attachments, craving or desire to the five aggregates, such a person experiences then transcends the very causes of suffering.

In this way, the insight wisdom or prajn of non-Self gives rise to cessation of suffering, and not an intellectual debate over whether a self exists or not.

It is by realizing (not merely understanding intellectually, but making real in one's experience) the three marks of conditioned existence that one develops prajn, which is the antidote to the ignorance that lies at the root of all suffering. From the "tathagatagarbha-Mahayana" perspective (which diverges from the Theravadin understanding of Buddhism), however, a further step is required if full Buddhahood is to be attained: not only seeing what is impermanent, suffering and non-Self in the samsaric sphere, but equally recognising that which is truly Eternal, Blissful, Self, and Pure in the transcendental realm—the realm of Mahaparinirvana.

Interpretations of the Three Marks by Various Schools

Some Buddhist traditions assert that Anatta pervades everything, and is not limited to personality, or soul. These traditions assert that Nirvana also has the quality of Anatta, but that Nirvana (by definition) is the cessation of Dukkha and Anicca.

In Nagarjuna's says:

There is not the slightest difference.

Between Samsara and Nirvana: This verse points us to an interesting stress between dukkha and nirvana, through an argument based in anatta. This specific stress can be seen to be the key to (and possibly source for the development of) the deity yogas of vajrayana.

The sutra path enjoins us to identify the entire world (internally and externally) as samsara—a continual churning of suffering that nobody wants to be part of. Our practice is that of leaving the shores of samsara.

On the other hand, we are told that unconditioned, enlightened activity is not actually different from samsara.

Whereas the deity yoga of vajrayana enjoins us to identify the entire world as nirvana—a continual play of enlightening activity that everyone wishes to be a part of. Our practice here is that of arriving at the shores of nirvana.

At this level, the distinction between Sutra and Vajrayana remain that of view (departing vs. arriving), but basically the practitioner remains involved in undergoing a transformative development to his or her Weltanschauung, and in this context, these practices remain rooted in psychological change, grounded in the development of Samatha, or training in concentration.

However, there are certain practices in Tantra which are not solely concerned with psychological change; these revolve around the basic idea that it is possible to induce deep levels of concentration through psycho-physical methods as a result of special exercises. The purpose remains the same (to achieve liberating view), but the method involves a 'short cut' for the training in Samatha.

Skandha

The five skandhas (Sanskrit) or five khandhas (Pli) are the five "aggregates" which constitute individual experience according to Buddhist phenomenology. Moreover, in Buddhism, a "person" is made up of the five skandhas, beyond which there is no "self."

In the Theravada tradition, suffering arises when one identifies with or otherwise clings to an aggregate; hence, suffering is extinguished by relinquishing attachment to aggregates. The Mahayana tradition further puts forth that ultimate freedom is realized by deeply penetrating the instrinsically empty nature of all aggregates.

Outside of Buddhist contexts, "skandha" can mean heap, bundle or tree trunk.

Definition: Buddhist doctrine describes one physical and four mental skandhas, which conveys the relative experience of the world by an individual. The five skandhas are:

1. "form" or "matter" (sa., pi. rkpa):includes both external and internal matter. Externally, rupa is the physical world. Internally, rupa includes the material body and the physical sense organs.
2. "sensation" or "feeling" (sa., pi. vedan):sensing pleasant, unpleasant or neutral (neither-pleasant-nor-unpleasant). Generally considered not to include "emotions."
3. "perception" or "cognition" (sa. samjn, pi. sann):registers whether sense data is recognized or not (for instance, the sound of a bell or the shape of a tree). From samyutta-nana, conditioned knowledge. It is ordinarily conditioned by ones past sankhara, and therefore conveys a coloured image of reality.
4. "mental formations" or "volition" (sa. samskra, pi. sankhra) :all types of mental habits, thoughts, ideas, opinions, compulsions, and decisions. Sankhra arise from contact with form or from other mental factors. Sankhra are the source of karma.

5. "consciousness" (sa. vijnna, pi. vinnma):conscious base that support all experience. Alternately, the Abhidhamma envisions consciousness as series of rapidly changing interconnected discrete acts of cognizance.

See examples of definitional references to the aggregates in Buddhist primary sources.

The aggregates are causally related as follows:

- The four "Great Elements" (mahbhkta: earth, wind, fire and water) give rise to form (rkpa).
- Through the contact (phassa) of an external object, a sense organ and consciousness (vinnma), form (rkpa) gives rise to the three "name" (nma) aggregates of feeling, perception and mental formation. (The nma aggregates are caused by either contact with an object or the arising of concomitant mental factors.)

Bhikkhu Bodhi (2000b, p. 840) states that an examination of the aggregates has a "critical role" in the Buddha's teaching for multiple reasons, including:

1. *Understanding the Four Noble Truths:* The five aggregates are the "ultimate referent" in the Buddha's elaboration on suffering (dukkha) in his First Noble Truth (see excerpted quote below) and "since all four truths revolve around suffering, understanding the aggregates is essential for understanding the Four Noble Truths as a whole."
2. *Future Suffering's Cause:* The five aggregates are the substrata for clinging and thus "contribute to the causal origination of future suffering."
3. *Release:* Clinging must be removed from the five aggregates in order to achieve release.

Below, excerpts from the Pli literature will bear out Bhikkhu Bodhi's assessment.

Suffering's Ultimate Referent: In the Buddha's first discourse, the "Dhammacakkappavattana Sutta" ("The Setting in Motion the Wheel of Truth Discourse," SN 56:11), he provides a classic elaboration on the first of his Four Noble Truths, "The Truth of Suffering" (Dukkhasacca):

> *"The Noble Truth of Suffering (dukkha), monks, is this: Birth is suffering, aging is suffering, sickness is suffering, death is suffering, association with the unpleasant is suffering, dissociation from the pleasant is suffering, not to receive what one desires is suffering — in brief the five aggregates subject to grasping are suffering." [Boldface added.] (Trans. from the Pali by Piyadassi Thera, 1999.)*

According to Thanissaro (2002)

> *"Prior to the Buddha, the Pali word khandha had very ordinary meanings: A khandha could be a pile, a bundle, a heap, a mass. It could also be the trunk of a tree. In his first sermon, though, the Buddha gave it a new, psychological meaning, introducing the term 'clinging-khandhas' to summarize his analysis of the truth of stress and suffering. Throughout the remainder of his teaching career, he referred to these psychological khandhas time and again."*

In what way are the aggregates suffering? For this we can turn to Khandhavagga suttas.

Future Suffering's Cause: The Samyutta Nikaya contains a book entitled the "Khandhavagga" ("The Book of Aggregates") compiling over a hundred suttas related to the five aggregates. Typical of these suttas is the "Upadaparitassana Sutta" ("Agitation through Clinging Discourse," SN 22:7), which states in part:

> *The instructed noble disciple... does not regard form [or other aggregates] as self, or self as possessing form, or form as in self, or self as in form. That form of his changes and alters. Despite the change and alteration of form, his consciousness does not become preoccupied with the change of form.... [T]hrough non-clinging he does not become agitated." (Trans. by Bodhi, 2000b, pp. 865-866.)*

Put another way, if we were to self-identify with an aggregate then we would cling (upadana) to such; and, given that all aggregates are impermanent (anicca), it would then be likely that at some level we would experience agitation (paritassati) or loss or grief or stress or suffering (see dukkha). Therefore, if we want to be free of suffering, it is wise to experience the aggregates clearly, without clinging or craving (tanha), as apart from any notion of self (anatta). Many of the suttas in the Khandhavagga express the aggregates in the context of the following sequence:

I. An uninstructed worldling (assutav puthujjana)
   1. regards: form as self; self as possessing form; form as in self; self as in form.
   2. lives obsessed by the notions: I am form; and/or, form is mine
   3. this form changes
   4. with the changes of form, there arises dukkha

II. An instructed noble disciple (sutav ariyasvaka) does not regard form as self, etc., and thus, when form changes, dukkha does not arise. (Note that, in each of the suttas where the above forumla is used, subsequent verses replace "form" with each of the other aggregates: sensation, perception, volition and consciousness.)

Example of Aggregate-Clinging to give a simplistic example, if one believes "this body is mine" or "I exist within this body," then as their body ages, becomes ill and approaches death, such a person will likely experience longing for youth or health or eternal life, will likely dread aging and sickness and death, and will likely spend much time and energy lost in fears and fantasies.In the Nikayas, such is likened to shooting oneself with a second arrow, where the first arrow is a physical phenomenon (such as, in this case, a bodily manifestation associated with aging or illness or dying) and the second is the mental anguish of the undisciplined mind associated with the physical phenomenon (see the Sallatha Sutta).On the other hand, one with a disciplined mind who is able to see this body as a set of aggregates will be free of such fear, frustration and time-consuming escapism.

But how does one become aware of and then let go of ones own identification with or clinging to the aggregates? Below is an excerpt from the classic Satipatthana Sutta that shows how traditional mindfulness practices can awaken understanding, release and wisdom.

Release through Aggregate-contemplation: In the classic Theravada meditation reference, the "Satipammhna Sutta" ("The Foundations of Mindfulness Discourse," MN 10), the Buddha provides four bases for establishing mindfulness: body (kaya), sensations (vedana), mind (citta) and mental objects (dhamma).

When discussing mental objects as a basis for meditation, the Buddha identifies five objects, including the aggregates. Regarding meditation on the aggregates, the Buddha states:

> *"How, monks, does a monk live contemplating mental objects in the mental objects of the five aggregates of clinging?*
>
> *"Herein, monks, a monk thinks, 'Thus is material form; thus is the arising of material form; and thus is the disappearance of material form. Thus is feeling; thus is the arising of feeling; and thus is the disappearance of feeling. Thus is perception; thus is the arising of perception; and thus is the disappearance of perception. Thus are formations; thus is the arising of formations; and thus is the disappearance of formations. Thus is consciousness; thus is the arising of consciousness; and thus is the disappearance of consciousness.'*
>
> *"Or his mindfulness is established with the thought, 'Mental objects exist,' to the extent necessary just for knowledge and mindfulness, and he lives detached, and clings to nothing in the world. Thus also, monks, a monk lives contemplating mental objects in the mental objects of the five aggregates of clinging." (Nyanasatta, trans., 1994.)*

Thus, through mindfulness contemplation, one sees an "aggregate as an aggregate"—sees it arising and dissipating. Such clear seeing creates a space between the aggregate and clinging, a space that will prevent or enervate the arising and propagation of clinging, thereby diminishing future suffering.

As clinging disappears, so too empty notions of self. In the Mahasunnata Sutta ("The Greater Discourse on Emptiness," MN 122), after reiterating the aforementioned aggregate-contemplation instructions (for instance, "Thus is form; thus is the arising of form; and, thus is the disappearance of form"), the Buddha states:

> *"When he [a monk] abides contemplating rise and fall in these five aggregates affected by clinging, the conceit 'I am' based on these five aggregates affected by clinging is abandoned in him...." (Nanamoli & Bodhi, 2001, p. 975.)*

In a complementary fashion, in the Buddha's second discourse, the Anattalakkhana Sutta ("The Characteristic of Nonself," SN 22:59), the Buddha instructs:

> *"Monks, form is nonself. For if, monks, form were self, this form would not lead to affliction, and it would be possible to [manipulate] form [in the following manner]: 'Let my form be thus; let my form not be thus....' [Identical statements are made regarding feeling, perception, volitional formations and consciousness.]*
>
> *"Seeing thus [for instance, through contemplation], monks, the instructed noble disciple becomes disenchanted with form [and the other aggregates].... Being disenchanted, he becomes dispassionate. Through dispassion [his mind] is liberated." (Bodhi, 2005, pp. 341-2.)*

As seen below, the Mahayana tradition continues this use of the aggregates to achieve self-liberation.

Mahayana Perspectives: In one of Mahayana Buddhism's most famous declarations, the aggregates are referenced:

> *"Form is emptiness, emptiness is form."*

What does this mean? To what degree is it a departure from the aforementioned Theravada perspective? Moreover, more generally, how are the aggregates used in the Mahayana literature? These questions are addressed below.

The intrinsic emptiness of all things

The classic "Prajnaparamita Hridaya Sutra" ("Heart Sutra") begins:

The Bodhisattva Avalokita,

while moving in the deep course of Perfect Understanding,

shed light on the five skandhas

and found them equally empty [of self].

After this penetration, he overcame all pain.

From its very first verse, the Heart Sutra introduces an alternative practice and worldview to the Theravada perspective of the aggregates:

- *Prajnaparamita:* Whereas Theravada meditation practices with the aggregates generally use change-penetrating vipassana meditation, here the non-dualistic prajnaparamita practice is invoked.
- *Svabhava:* In Theravada texts, when "emptiness of self" is mentioned, the English word "self" is a translation of the Pali word "atta" (Sanskrit, "atman"); in the Heart Sutra, the English word "self" is a translation of the Sanskrit word "sva-bhava". According to Red Pine, "The 'self' (sva)... was more generalized in its application than 'ego' (atman) and referred not only to beings but to any inherent substance that could be identified as existing in time or space as a permanent or independent entity." (Italics added.)

In other words, whereas the Sutta Pitaka typically instructs one to apprehend the aggregates without clinging or self-identification, Prajnaparamita leads one to apprehend the aggregates as having no intrinsic reality.

In the Heart Sutra's second verse, after rising from his aggregate meditation, Avalokiteshvara declares:

"Form is emptiness, emptiness is form,

form does not differ from emptiness, emptiness does not differ from form.

The same is true with feelings, perceptions, mental formations and consciousness."

Red Pine Comments:

"That form is empty was one of the Buddha's earliest and most frequent pronouncements. But in the light of Prajnaparamita, form is not simply empty, it is so completely empty, it is emptiness itself, which turns out to be the same as form itself.... All separations are delusions. But if each of the skandhas is one with emptiness, and emptiness is one with each of the skandhas, then everything occupies the same indivisible space, which is emptiness.... Everything is empty, and empty is everything.

D.T. Suzuki further states:

"When the sutra says that the five Skandhas have the character of emptiness..., the sense is: no limiting qualities are to be attributed to the Absolute; while it is immanent in all concrete and particular objects, it is not in itself definable."

We can see in sutra form, an expounding of the Buddha's doctrine of Anatman (No-self, No-I). Here the doctrine of the Five Skandha supports the concept of Anatman. For if there is "No-self", "No-I", then what experiences? The answer lies in the five skandha.

The Truth of Our Insubstantiality: According to Chogyam Trungpa (1976, pp. 20-22), the five skandhas are "a set of Buddhist concepts which describe experience as a five-step process" and that "the whole development of the five skandhas...is an attempt on our part to shield ourselves from the truth of our insubstantiality," while "the practice of meditation is to see the transparency of this shield." (ibid, p.23)

Relation to Other Buddhist Concepts

Other fundamental Buddhist concepts associated with the five skandhas include:

- *Samsara:*It is through the five skandhas that the world (samsara) is experienced, and nothing is experienced apart from the five skandhas.

- *Three Characteristics:* It is through the five skandhas that impermanence (anicca) is experienced, that suffering (duhkha) arises, and that "non-self" (anatta or anatman) can be realized.
- *Four Paramatthas:*The Abhidhamma and post-canonical Pali texts create a meta-scheme for the Sutta Pitaka's conceptions of aggregates, sense bases and elements. This meta-scheme is known as the four paramatthas or four ultimate realities:
- consciousness
- mental factors
- material phenomena
- Nibbna The mapping between the ultimate realities and the aggregates is represented in the chart to the right.
- Twelve Nidanas / Dependent Origination:

It is through the five skandhas that clinging (upadana) occurs, a pivotal link in the endless chain of suffering.

- *Eighteen Dhatus :*The eighteen dhatus function through the five aggregates. The eighteen dhatus can be arranged into six triads, where each triad is composed of a sense organ, a sense object and sense consciousness. In regards to the aggregates :
- The first five sense organs (eye, ear, nose, tongue, body) are derivates of form. The sixth sense organ (mind) is part of consciousness.
- The first five sense objects (visible forms, sound, smell, taste, touch) are also derivatives of form. The sixth sense object (mental object) includes form, sensation, perception and mental formations.
- The six sense consciousness are the basis for consciousness.

References in Buddhist Literature

The table below briefly cites Buddhist primary sources that characterize different aspects of the aggregates. This table is by no means exhaustive.

Buddhist Cosmology the Canonical Buddhist Scriptures and Commentaries

The self-consistent Buddhist cosmology which is presented in commentaries and works of Abhidharma in both Theravda and Mahyna traditions, is the end-product of an analysis and reconciliation of cosmological comments found in the Buddhist sktra and vinaya traditions. No single sktra sets out the entire structure of the universe. However, in several sktras the Buddha describes other worlds and states of being, and other sktras describe the origin and destruction of the universe. The synthesis of these data into a single comprehensive system must have taken place early in the history of Buddhism, as the system described in the Pli Vibhajyavda tradition (represented by today's Theravdins) agrees, despite some trivial inconsistencies of nomenclature, with the Sarvstivda tradition which is preserved by Mahyna Buddhists.

The picture of the world presented in Buddhist cosmological descriptions cannot be taken as a literal description of the shape of the universe. It is inconsistent, and cannot be made consistent, with astronomical data that were already known in ancient India. However, it is not intended to be a description of how humans perceive their world; rather, it is the universe as seen through the divyacakcus (Pli: dibbacakkhu), the "divine eye" by which a Buddha or an arhat who has cultivated this faculty can perceive all of the other worlds and the beings arising (being born) and passing away (dying) within them, and can tell from what state they have been reborn and into what state they will be reborn.

Buddhist cosmology can accordingly be divided into two related kinds: spatial cosmology, which describes the arrangement of the various worlds within the universe, and temporal cosmology, which describes how those worlds come into existence, and how they pass away.

Spatial Cosmology

Spatial cosmology can also be divided into two branches. The vertical cosmology describes the arrangement of worlds in a vertical pattern, some being higher and some lower. The sahasra cosmology describes the grouping of these vertical worlds into sets of thousands, millions or billions.

Vertical Cosmology

In the vertical cosmology, the universe exists of many worlds (lok) – one might say "planes" – stacked one upon the next in layers. Each world corresponds to a mental state or a state of being. A world is not, however, a location so much as it is the beings which compose it; it is sustained by their karma and if the beings in a world all die or disappear, the world disappears too. Likewise, a world comes into existence when the first being is born into it. The physical separation is not so important as the difference in mental state; humans and animals, though they partially share the same physical environments, still belong to different worlds because their minds perceive and react to those environments differently.

The vertical cosmology is divided into three realms, or Tridhatu, each corresponding to a different type of mentality. These three (Tridhatu) are the Rupadhatu, the Rupadhatu, and the Kamadhatu. This technical division does not correspond to the more informal categorization of the "six realms". In the latter scheme, all of the beings born in the Rupadhatu and the Rupadhatu may be classified as "gods" or "deities" (dev), as can a considerable fraction of the beings born in the Kamadhatu, even though the deities of the Kamadhatu differ more from those of the Rupadhatu than they do from humans. It is to be understood that deva is an imprecise term referring to any being living in a longer-lived and generally more blissful state than humans. Most of them are not "gods" in the common sense of the term, having little or no concern with the human world and rarely if ever interacting with it; only the lowest deities of the Kamadhatu correspond to the gods described in many polytheistic religions.

The term "brahm" is used both as a name and as a generic term for one of the higher devas. In its broadest sense, it can refer to any of the inhabitants of the Rupadhatu and the Rupadhatu. In more restricted senses, it can refer to an inhabitant of one of the nine lower worlds of the Rupadhatu, or in its narrowest sense, to the three lowest worlds of the Rupadhatu. A large number of devas use the name "Brahm", *e.g.* Brahm Sahampati, Brahm Sanatkumra, Baka Brahm, etc. It is not always clear which world they belong to, although it must always be one of the worlds of the Rupadhatu below the Zuddhvsa worlds.

There are four types of Rupadhatu devas, corresponding to the four types of arupadhynas:

- *Naivasanjnnsanjnyatana or Nevasannnsannyatana:* "Sphere of neither perception nor non-perception". In this sphere the formless beings have gone beyond a mere negation of perception and have attained a liminal state where they do not engage in "perception" (sanna, recognition of particulars by their marks) but are not wholly unconscious. This was the sphere reached by Udraka Ramaputta (Pali: Uddaka Ramaputta), the second of the Buddha's two teachers, who considered it equivalent to enlightenment.
- *Kiccanyyatana or Kincannyatana:* "Sphere of Nothingness" (literally "lacking anything"). In this sphere formless beings dwell contemplating upon the thought that "there is no thing". This is considered a form of perception, though a very subtle one. This was the sphere reached by r.
- *Alarakalama:* the first of the Buddha's two teachers; he considered it to be equivalent to enlightenment.
- *Vijnnnantyyatana or Vinnnnancyatana:* "Sphere of Infinite Consciousness". In this sphere formless beings dwell meditating on their consciousness (vijnna) as infinitely pervasive.
- *Kenantyyatana or Kinancyatana:* "Sphere of Infinite Space". In this sphere formless beings dwell meditating upon space or extension (kla) as infinitely pervasive.

Rupadhatu

The Rupadhatu or "Form realm" is, as the name implies, the first of the physical realms; its inhabitants all have a location and bodies of a sort, though those bodies are composed of a subtle substance which is of itself invisible to the inhabitants of the Kamadhatu. According to the Janavasabha Sutta, when a brahma (a being from the Brahma-world of the Rupadhatu) wishes to visit a deva of the Tryastricea heaven (in the Kamadhatu), he has to assume a "grosser form" in order to be visible to them.

The beings of the Form realm are not subject to the extremes of pleasure and pain, or governed by desires for things pleasing to the senses, as the beings of the Kamadhatu are. The bodies of Form realm beings do not have sexual distinctions.

Like the beings of the Rupadhatu, the dwellers in the Rupadhatu have minds corresponding to the dhyanas (Pali: jhanas). In their case it is the four lower dhyanas or rkpadhyanas. However, although the beings of the Rupadhatu can be divided into four broad grades corresponding to these four dhyanas, each of them is subdivided into further grades, three for each of the four dhyanas and five for the Zuddhvsa devas, for a total of seventeen grades (the Theravda tradition counts one less grade in the highest dhayna for a total of sixteen).

Physically, the Rupadhatu consists of a series of planes stacked on top of each other, each one in a series of steps half the size of the previous one as one descends. In part, this reflects the fact that the devas are also thought of as physically larger on the higher planes. The highest planes are also broader in extent than the ones lower down, as discussed in the section on Sahasra cosmology. The height of these planes is expressed in yojanas, a measurement of very uncertain length, but sometimes taken to be about 4,000 times the height of a man, and so approximately 4.54 miles or 7.32 kilometers.

Zuddhvsa Worlds

The Zuddhvsa (Pali: Suddhvsa; Tib: gnas gtsang.ma) worlds, or "Pure Abodes", are distinct from the other worlds of the Rupadhatu in that they do not house beings who have been born there through ordinary merit or meditative attainments, but only those Anagmins ("Non-returners") who are already on the path to Arhat-hood and who will attain enlightenment directly from the Zuddhvsa worlds without being reborn in a lower plane (Anagmins can also be born on lower planes). Every Zuddhvsa deva is therefore a protector of Buddhism.

Because a Zuddhvsa deva will never be reborn outside the Zuddhvsa worlds, no Bodhisattva is ever born in these worlds, as a Bodhisattva must ultimately be reborn as a human being.

Since these devas rise from lower planes only due to the teaching of a Buddha, they can remain empty for very long periods if no Buddha arises. However, unlike the lower worlds, the Zuddhvsa worlds are never destroyed by natural catastrophe. The Zuddhvsa devas predict the coming of a Buddha and, taking the guise of brahmins, reveal to human beings the signs by which a Buddha can be recognized. They also ensure that a Bodhisattva in his last life will see the four signs that will lead to his renunciation.

The five Zuddhvsa worlds are:

- *Akanicmha or Akanimmha:* World of devas "equal in rank" (literally: having no one as the youngest). The highest of all the Rupadhatu words, it is often used to refer to the highest extreme of the universe. The current Zakra will eventually be born there. The duration of life in Akanimha is 16,000 kalpas (Vibhajyavada tradition). The height of this world is 167,772,160 yojanas above the Earth.
- *Sudarana or Sudass:* The "clear-seeing" devas live in a world similar to and friendly with the Akanimha world. The height of this world is 83,886,080 yojanas above the Earth.
- *Suda or Sudassa:* The world of the "beautiful" devas are said to be the place of rebirth for five kinds of Anagmins. The height of this world is 41,943,040 yojanas above the Earth.
- *Atapa or Atappa:* The world of the "untroubled" devas, whose company those of lower realms wish for. The height of this world is 20,971,520 yojanas above the Earth.
- *Avaha or Aviha:* The world of the "not falling" devas, perhaps the most common destination for reborn Anagmins. Many achieve arhatship directly in this world, but some pass away and are reborn in sequentially

higher worlds of the Pure Abodes until they are at last reborn in the Akanimha world. These are called in Pali uddhansotas, "those whose stream goes upward". The duration of life in Av[ha is 1,000 kalpas (Vibhajyavada tradition). The height of this world is 10,485,760 yojanas above the Earth.

Bhatphala Worlds

The mental state of the devas of the Behatphala worlds corresponds to the fourth dhyana, and is characterized by equanimity (upekkha). The Behatphala worlds form the upper limit to the destruction of the universe by wind at the end of a mahakalpa (see Temporal cosmology below), that is, they are spared such destruction.

- *Asannasatta (Sanskrit: Asacjnasattva) (Vibhajyaada tradition only):* "Unconscious beings", devas who have attained a high dhyana (similar to that of the Formless Realm), and, wishing to avoid the perils of perception, have achieved a state of non-perception in which they endure for a time. After a while, however, perception arises again and they fall into a lower state.
- *Behatphala or Vehapphala:* Devas "having great fruit". Their lifespan is 500 mahakalpas. (Vibhajyavada tradition). Some Anagmins are reborn here. The height of this world is 5,242,880 yojanas above the Earth.
- *Punyaprasava:* The world of the devas who are the "offspring of merit". The height of this world is 2,621,440 yojanas above the Earth.
- *Anabhraka:* The world of the "cloudless" devas. The height of this world is 1,310,720 yojanas above the Earth.

Zubhaktsna Worlds

The mental state of the devas of the Zubhaketsna worlds corresponds to the third dhyana, and is characterized by a quiet joy (sukha). These devas have bodies that radiate a steady light. The Zubhaketsna worlds form the upper limit to the destruction of the universe by water at the end of a mahakalpa (see Temporal cosmology below), that is, the flood of water does not rise high enough to reach them.

- *Zubhaktsna or Subhakinna / Subhakinha:* The world of devas of "total beauty". Their lifespan is 64 mahakalpas (some sources: 4 mahakalpas) according to the Vibhajyavada tradition. 64 mahakalpas is the interval between destructions of the universe by wind, including the Zubhaketsna worlds. The height of this world is 655,360 yojanas above the Earth.
- *Apramnaeubha or Appamnasubha:* The world of devas of "limitless beauty". Their lifespan is 32 mahakalpas (Vibhajyavada tradition). They possess "faith, virtue, learning, munificence and wisdom". The height of this world is 327,680 yojanas above the Earth.
- *Parittaeubha or Parittasubha:* The world of devas of "limited beauty". Their lifespan is 16 mahakalpas. The height of this world is 163,840 yojanas above the Earth.

Bhasvara Worlds

The mental state of the devas of the bhasvara worlds corresponds to the second dhyana, and is characterized by delight (priti) as well as joy (sukha); the bhasvara devas are said to shout aloud in their joy, crying aho sukham! ("Oh joy!"). These devas have bodies that emit flashing rays of light like lightning. They are said to have similar bodies (to each other) but diverse perceptions.

The bhasvara worlds form the upper limit to the destruction of the universe by fire at the end of a mahakalpa (see Temporal cosmology below), that is, the column of fire does not rise high enough to reach them. After the destruction of the world, at the beginning of the vivartakalpa, the worlds are first populated by beings reborn from the Bhasvara worlds.

- *Bhasvara or Bhassara:* The world of devas "possessing splendor". The lifespan of the bhasvara devas is 8 mahakalpas (others: 2 mahakalpas). Eight mahakalpas is the interval between destructions of the universal by water, which includes the bhasvara worlds. The height of this world is 81,920 yojanas above the Earth.

- *Apramnbha or Appamnbha:* The world of devas of "limitless light", a concept on which they meditate. Their lifespan is 4 mahakalpas. The height of this world is 40,960 yojanas above the Earth.
- *Parittbha or Parittbha:* The world of devas of "limited light". Their lifespan is 2 mahakalpas. The height of this world is 20,480 yojanas above the Earth.

Brahm Worlds

The mental state of the devas of the Brahm worlds corresponds to the first dhyana, and is characterized by observation (vitarka) and reflection (vicra) as well as delight (priti) and joy (sukha). The Brahm worlds, together with the other lower worlds of the universe, are destroyed by fire at the end of a mahakalpa (see Temporal cosmology below).

- *Mahabrahm:* The world of "Great Brahm", believed by many to be the creator of the world, and having as his titles "Brahm, Great Brahm, the Conqueror, the Unconquered, the All-Seeing, All-Powerful, the Lord, the Maker and Creator, the Ruler, Appointer and Orderer, Father of All That Have Been and Shall Be." According to the Brahmajla Sutta (DN.1), a Mahabrahm is a being from the bhasvara worlds who falls into a lower world through exhaustion of his merits and is reborn alone in the Brahma-world; forgetting his former existence, he imagines himself to have come into existence without cause. Note that even such a high-ranking deity has no intrinsic knowledge of the worlds above his own. Mahabrahm is 1^1D_2 yojanas tall. His lifespan variously said to be 1 kalpa (Vibhajyavada tradition) or 1^1D_2 kalpas long (Sarvstivda tradition), although it would seem that it could be no longer than $^3D_4$ of a mahakalpa, *i.e.*, all of the mahakalpa except for the Sacvartasthyikalpa, because that is the total length of time between the rebuilding of the lower world and its destruction. It is unclear what period of time "kalpa" refers to in this case. The height of this world is 10,240 yojanas above the Earth.

- *Brahmapurohita:* The "Ministers of Brahm" are beings, also originally from the bhasvara worlds, that are born as companions to Mahabrahm after he has spent some time alone. Since they arise subsequent to his thought of a desire for companions, he believes himself to be their creator, and they likewise believe him to be their creator and lord. They are 1 yojana in height and their lifespan is variously said to be $^{1}D_{2}$ of a kalpa (Vibhajyavada tradition) or a whole kalpa (Sarvstivda tradition). If they are later reborn in a lower world, and come to recall some part of their last existence, they teach the doctrine of Brahm as creator as a revealed truth. The height of this world is 5,120 yojanas above the Earth.
- *Brahmapricadya or Brahmaprisajja:* The "Councilors of Brahm" or the devas "belonging to the assembly of Brahm". They are also called Brahmakyika, but this name can be used for any of the inhabitants of the Brahma-worlds. They are half a yojana in height and their lifespan is variously said to be $^{1}D_{3}$ of a kalpa (Vibhajyavada tradition) or $^{1}D_{2}$ of a kalpa (Sarvstivda tradition). The height of this world is 2,560 yojanas above the Earth.

Kamadhatu

The beings born in the Kamadhatu (Pali: Kamaloka; Tib: 'dod.pa'i khams) differ in degree of happiness, but they are all, other than arhats and Buddhas, under the domination of Mara and are bound by desire, which causes them suffering.

Heavens

The following four worlds are bounded planes. each 80,000 yojanas square, which float in the air above the top of Mount Sumeru. Although all of the worlds inhabited by devas (that is, all the worlds down to the Caturmaharajikakayika world and sometimes including the Asuras) are sometimes called "heavens", in the western sense of the word the term best applies to the four worlds listed below:

- *Parinirmita-vasavartin or Paranimmita-vasavatti:* The heaven of devas "with power over (others') creations". These devas do not create pleasing forms that they desire for themselves, but their desires are fulfilled by the acts of other devas who wish for their favor. The ruler of this world is called Va[avartin (Pali: Vasavatti), who has longer life, greater beauty, more power and happiness and more delightful sense-objects than the other devas of his world. This world is also the home of the devaputra (being of divine race) called Mra, who endeavors to keep all beings of the Kamadhatu in the grip of sensual pleasures. Mra is also sometimes called Vavartin, but in general these two dwellers in this world are kept distinct. The beings of this world are 4,500 feet tall and live for 9,216,000,000 years (Sarvstivda tradition). The height of this world is 1,280 yojanas above the Earth.
- *Nirmnarati or Nimmnarati:* The world of devas "delighting in their creations". The devas of this world are capable of making any appearance to please themselves. The lord of this world is called Sunirmita (Pali Sunimmita); his wife is the rebirth of Viskh, formerly the chief of the upsiks (female lay devotees) of the Buddha. The beings of this world are 3,750 feet tall and live for 2,304,000,000 years (Sarvstivda tradition). The height of this world is 640 yojanas above the Earth.
- *Tucita or Tusita:* The world of the "joyful" devas. This world is best known for being the world in which a Bodhisattva lives before being reborn in the world of humans. Until a few thousand years ago, the Bodhisattva of this world was Zvetaketu (Pali: Setaketu), who was reborn as Siddhrtha, who would become the Buddha Zkyamuni; since then the Bodhisattva has been Ntha (or Nthadeva) who will be reborn as Ajita and will become the Buddha Maitreya (Pali Metteyya). While this Bodhisattva is the foremost of the dwellers in Tucita, the ruler of this world is another deva called Santucita (Pali: Santusita). The beings of this world are

3,000 feet tall and live for 576,000,000 years (Sarvstivda tradition). The height of this world is 320 yojanas above the Earth.

- *Yama:* Sometimes called the "heaven without fighting", because it is the lowest of the heavens to be physically separated from the tumults of the earthly world. Its ruler is the deva Suyma; according to some, his wife is the rebirth of Sirim, a courtesan of Rjag[ha in the Buddha's time who was generous to the monks. The beings of this world are 2,250 feet tall and live for 144,000,000 years (Sarvstivda tradition). The height of this world is 160 yojanas above the Earth.

Worlds of Sumeru

The world-mountain of Sumeru is an immense, strangely shaped peak which arises in the center of the world, and around which the Sun and Moon revolve. Its base rests in a vast ocean, and it is surrounded by several rings of lesser mountain ranges and oceans. The three worlds listed below are all located on or around Sumeru: the Tryastricea devas live on its peak, the Cturmahrjikakyika devas live on its slopes, and the Asuras live in the ocean at its base. Sumeru and its surrounding oceans and mountains are the home not just of these deities, but also vast assemblies of fantastic fairy-tale creatures who only rarely intrude on the human world.

Earthly Realms

- *Manucyaloka:* This is the world of humans and human-like beings who live on the surface of the earth. The mountain-rings that engird Sumeru are surrounded by a vast ocean, which fills most of the world. The ocean is in turn surrounded by a circular mountain wall called Cakrav
- A which marks the horizontal limit of the world. In this ocean there are four continents which are, relatively speaking, small islands in it. Because of the immenseness of the ocean, they cannot be reached from each other by ordinary sailing vessels, although in the past, when the cakravartin kings ruled, communication

between the continents was possible by means of the treasure called the cakraratna (Pli cakkaratana), which a cakravartin and his retinue could use to fly through the air between the continents. The four continents are:

- Jambudvipa or Jambudipa is located in the south and is the dwelling of ordinary human beings. It is said to be shaped "like a cart", or rather a blunt-nosed triangle with the point facing south. (This description probably echoes the shape of the coastline of southern India.) It is 10,000 yojanas in extent (Vibhajyavda tradition) or has a perimeter of 6,000 yojanas (Sarvstivda tradition) to which can be added the southern coast of only 3 ^{1}D $_{2}$ yojanas' length. The continent takes its name from a giant Jambu tree (Syzygium cumini), 100 yojanas tall, which grows in the middle of the continent. Every continent has one of these giant trees. All Buddhas appear in Jambudvipa. The people here are five to six feet tall and their length of life varies between 80,000 and 10 years.
- Pkrvavideha or Pubbavideha is located in the east, and is shaped like a semicircle with the flat side pointing westward (*i.e.*, towards Sumeru). It is 7,000 yojanas in extent (Vibhajyavda tradition) or has a perimeter of 6,350 yojanas of which the flat side is 2,000 yojanas long (Sarvstivda tradition). Its tree is the acacia. The people here are about 12 feet tall and they live for 250 years.
- Aparagodniya or Aparagoyna is located in the west, and is shaped like a circle with a circumference of about 7,500 yojanas (Sarvstivda tradition). The tree of this continent is a giant Kadambu tree. The human inhabitants of this continent do not live in houses but sleep on the ground. They are about 24 feet tall and they live for 500 years.
- Uttarakuru is located in the north, and is shaped like a square. It has a perimter of 8,000 yojanas, being 2,000 yojanas on each side. This continent's tree is called a

kalpavskca or kalpa-tree, because it lasts for the entire kalpa. The inhabitants of Uttarakuru are said to be extraordinarily wealthy. They do not need to labour for a living, as their food grows by itself, and they have no private property. They have cities built in the air. They are about 48 feet tall and live for 1,000 years, and they are under the protection of Vaisravaga.

- *Tiryagyoni-loka or Tiracchna-yoni:* This world comprises all members of the animal kingdom that are capable of feeling suffering, from the smallest insect to the elephant.
- *Pretaloka or Petaloka:* The pretas, or "hungry ghosts", are mostly dwellers on earth, though due to their mental state they perceive it very differently from humans. They live for the most part in desert and waste places.

Narakas

Naraka or Niraya is the name given to one of the worlds of greatest suffering, usually translated into English as "hell" or "purgatory". As with the other realms, a being is born into one of these worlds as a result of his karma, and resides there for a finite length of time until his karma has achieved its full result, after which he will be reborn in one of the higher worlds as the result of an earlier karma that had not yet ripened. The mentality of a being in the hells corresponds to states of extreme fear and helpless anguish in humans.

Physically, Naraka is thought of as a series of layers extending below Jambudvipa into the earth. There are several schemes for counting these Narakas and enumerating their torments. One of the more common is that of the Eight Cold Narakas and Eight Hot Narakas.

Cold Narakas

- Arbuda – the "blister" Naraka
- Nirarbuda – the "burst blister" Naraka
- Amama – the Naraka of shivering
- Hahava – the Naraka of lamentation
- Huhuva – the Naraka of chattering teeth

- Utpala – the "blue lotus" Naraka
- Padma – the "lotus" Naraka
- Mahpadma – the "great lotus" Naraka

The Foundations of the Earth: All of the structures of the earth, Sumeru and the rest, extend downward to a depth of 80,000 yojanas below sea level – the same as the height of Sumeru above sea level. Below this is a layer of "golden earth", a substance compact and firm enough to support the weight of Sumeru. It is 320,000 yojanas in depth and so extends to 400,000 yojanas below sea level. The layer of golden earth in turn rests upon a layer of water, which is 8,000,000 yojanas in depth, going down to 8,400,000 yojanas below sea level. Below the layer of water is a "circle of wind", which is 16,000,000 yojanas in depth and also much broader in extent, supporting 1,000 different worlds upon it.

Sahasra Cosmology

While the vertical cosmology describes the arrangement of the worlds vertically, the sahasra (Sanskrit: "thousand") cosmology describes how they are grouped horizontally. The four heavens of the Kamadhatu, as mentioned, occupy a limited space no bigger than the top of Mount Sumeru. The three Brahm-worlds, however, stretch out as far as the mountain-wall of Cakrav-a, filling the entire sky. This whole group of worlds, from Mahabrahm down to the foundations of water, constitutes a single world-system. It corresponds to the extent of the universe that is destroyed by fire at the end of one mahakalpa.

Above Mahabrahm are the bhasvara worlds. These are not only higher but also wider in extent; they cover 1,000 separate world-systems, each with its own Sumeru, Cakrav a, Sun, Moon, and four continents. This system of 1,000 worlds is called a shasra-ck-ika-lokadhtu, or "small chiliocosm". It corresponds to the extent of the universe that is destroyed by water at the end of 8 mahakalpas.

Above the bhasvara worlds are the Zubhaketsna worlds, which cover 1,000 chiliocosms, or 1,000,000 world-systems. This larger system is called a dvishasra-madhyama-lokadhtu, or "medium dichiliocosm". It corresponds to the extent of the universe that is destroyed by wind at the end of 64 mahakalpas.

Likewise, above the Zubhaketsna worlds, the Zuddhvsa and Behatphala worlds cover 1,000 dichiliocosms, or 1,000,000,000 world-systems. This largest grouping is called a trishasra-mahshasra-lokadhtu or "great trichiliocosm".

Temporal Cosmology

Buddhist temporal cosmology describes how the universe comes into being and is dissolved. Like other Indian cosmologies, it assumes an infinite span of time and is cyclical. This does not mean that the same events occur in identical form with each cycle, but merely that, as with the cycles of day and night or summer and winter, certain natural events occur over and over to give some structure to time.

The basic unit of time measurement is the mahakalpa or "Great Eon". The exact length of this time in human years is never defined exactly, but it is meant to be very long, to be measured in billions of years if not longer. A mahakalpa is divided into four kalpas or "eons", each distinguished from the others by the stage of evolution of the universe during that kalpa. The four kalpas are:

- Vivartakalpa "Eon of evolution" – during this kalpa the universe comes into existence.
- Vivartasthyikalpa "Eon of evolution-duration" – during this kalpa the universe remains in existence in a steady state.
- Sacvartakalpa "Eon of dissolution" – during this kalpa the universe dissolves.
- Sacvartasthyikalpa "Eon of dissolution-duration" – during this kalpa the universe remains in a state of emptiness.

Each one of these kalpas is divided into twenty antarakalpas (Pali antarakappa, "inside eons") each of about the same length. For the Sacvartasthyikalpa this division is merely nominal, as nothing changes from one antarakalpa to the next; but for the other three kalpas it marks an interior cycle within the kalpa.

Vivartakalpa

The Vivartakalpa begins with the arising of the primordial wind, which begins the process of building up the structures of the universe that had been destroyed at the end of the last mahkalpa. As the extent of the destruction can vary, the nature of this evolution can vary as well, but it always takes the form of beings from a higher world being born into a lower world. The example of a Mahbrahm being the rebirth of a deceased bhsvara deva is just one instance of this, which continues throughout the Vivartakalpa until all the worlds are filled from the Brahmaloka down to Naraka. During the Vivartakalpa the first humans appear; they are not like present-day humans, but are beings shining in their own light, capable of moving through the air without mechanical aid, living for a very long time, and not requiring sustenance; they are more like a type of lower deity than present-day humans are.

Over time, they acquire a taste for physical nutriment, and as they consume it, their bodies become heavier and more like human bodies; they lose their ability to shine, and begin to acquire differences in their appearance, and their length of life decreases. They differentiate into two sexes and begin to become sexually active. Then greed, theft and violence arise among them, and they establish social distinctions and government and elect a king to rule them, called Mahsammata, "the great appointed one". Some of them begin to hunt and eat the flesh of animals, which have by now come into existence. These developments are described in the Agganna Sutta (DN.27).

Vivartasthyikalpa

The Vivartasthyikalpa begins when the first being is born into Naraka, thus filling the entire universe with beings. During the first antarakalpa of this eon, human lives are declining from a vast but unspecified number of years (but at least several tens of thousands of years) toward the modern lifespan of less than 100 years. At the beginning of the antarakalpa, people are still generally happy. They live under the rule of a universal monarch or "wheel-turning king" (cakravartin), who conquer. The Mahsudassana-sutta (DN.17) tells of the life of a cakravartin king, Mahsudassana (Sanskrit: Mahsudar[ana) who lived for 336,000 years. The Cakkavatti-sihanda-sutta (DN.26) tells of a later dynasty of cakravartins, Dahanemi (Sanskrit: Dhanemi) and five of his descendants, who had a lifespan of over 80,000 years. The seventh of this line of cakravartins broke with the traditions of his forefathers, refusing to abdicate his position at a certain age, pass the throne on to his son, and enter the life of a [ramana. As a result of his subsequent misrule, poverty increased; as a result of poverty, theft began; as a result of theft, capital punishment was instituted; and as a result of this contempt for life, murders and other crimes became rampant.

The human lifespan now quickly decreased from 80,000 to 100 years, apparently decreasing by about half with each generation (this is perhaps not to be taken literally), while with each generation other crimes and evils increased: lying, adultery, evil speech, greed and hatred, wrong views, incest and other sorts of sexual abnormalities, disrespect for parents and elders. During this period, according to the Mahpadna-sutta (DN.14) three of the four Buddhas of this antarakalpa lived: Krakucchanda Buddha (Pali: Kakusandha), at the time when the lifespan was 40,000 years; Kanakamuni Buddha (Pali: Kongamana) when the lifespan was 30,000 years; and K[yapa Buddha (Pali: Kassapa) when the lifespan was 20,000 years.

Our present time is taken to be toward the end of the first antarakalpa of this Vivartasthyikalpa, when the lifespan is less than 100 years, after the life of Zkyamuni Buddha (Pali: Sakyamuni), who lived to the age of 80.

The remainder of the antarakalpa is prophesied to be miserable: lifespans will continue to decrease, and all the evil tendencies of the past will reach their ultimate in destructiveness. People will live no longer than ten years, and will marry at five; foods will be poor and tasteless; no form of morality will be acknowledged. The most contemptuous and hateful people will become the rulers. Incest will be rampant. Hatred between people, even members of the same family, will grow until people think of each other as hunters do of their prey.

Eventually a great war will ensue, in which the most hostile and aggressive will arm themselves and go out to kill each other. The less aggressive will hide in forests and other secret places while the war rages. This war marks the end of the first antarakalpa.

Second Antarakalpa

At the end of the war, the survivors will emerge from their hiding places and repent their evil habits. As they begin to do good, their lifespan increases, and the health and welfare of the human race will also increase with it. After a long time, the descendants of those with a 10-year lifespan will live for 80,000 years, and at that time there will be a cakravartin king named SaEkha.

During his reign, the current bodhisattva in the Tucita heaven will descend and be reborn under the name of Ajita. He will then be known by the name of Maitreya (Pali: Metteyya). After Maitreya's time, the world will again worsen, and the lifespan will gradually decrease from 80,000 years to 10 years again, each antarakalpa being separated from the next by devastating war, with peaks of high civilization and morality in the middle. After the 19th antarakalpa, the lifespan will increase to 80,000 and then not decrease, because the Vivartasthyikalpa will have come to an end.

Sacvartakalpa

The Sacvartakalpa begins when beings cease to be born in Naraka. This cessation of birth then proceeds in reverse order up the vertical cosmology, *i.e.* pretas then cease to be born, then animals, then humans, and so on up to the realms of the deities. When these worlds as far as the Brahmaloka are devoid of inhabitants, a great fire consumes the entire physical structure of the world. It burns all the worlds below the bhasvara worlds. When they are destroyed, the Sacvartasthyikalpa begins.

Sachvartasthyikalpa

There is nothing to say about the Sacvartasthyikalpa, since nothing happens in it below the bhasvara worlds. It ends when the primordial wind begins to blow and build the structure of the worlds up again.

Other Destructions: The destruction by fire is the normal type of destruction that occurs at the end of the Sacvartakalpa. But every eighth mahakalpa, after seven destructions by fire, there is a destruction by water. This is more devastating, as it eliminates not just the Brahma worlds but also the bhasvara worlds.

Every sixty-fourth mahakalpa, after 56 destructions by fire and 7 destructions by water, there is a destruction by wind. This is the most devastating of all, as it also destroys the Zubhaketsna worlds. The higher worlds are never destıoyed.

Buddhism Dharma: Dharma (Sanskrit) or Dhamma (Pali) in Buddhism has two primary meanings: the teachings of the Buddha which lead to enlightenment, and the constituent factors of the experienced world. In East Asia, the character for Dharma is Ol, pronounced fI in Mandarin and hM in Japanese. The Tibetan translation of this term is chos (cho).

Buddha's Teachings: For practicing Buddhists, references to "dharma" (in English often "the" Dharma) primarily mean the teachings of the Buddha – sometimes called Buddhadharma by other schools of thought. Some Buddhists object to this type of terminology on the grounds that the Dharma is the universal and singular law of nature and to call it Buddhadharma misleadingly suggests that other kinds of Dharma may exist. Buddha Dharma itself, however, means "Path of Awakening" and so is a universal understanding of Dharma.

The status of the Dharma is regarded variably by different traditions. Some regard it as an ultimate and transcendent truth which is utterly beyond worldly things, somewhat like the Christian logos. Others, who regard the Buddha as simply an enlightened human being, see the Dharma as the 84,000 different teachings (the Kangyur/bka.'gyur) that the Buddha gave to various types of people based on their needs. The teachings are expedient means of raising doubt in the hearer's own cherished beliefs and view of life; when doubt has opened the door to the Truth, the teaching can be put aside.

"Dharma" usually refers inclusively not just to the sayings of the Buddha but to the later traditions of interpretation and addition that the various schools of Buddhism have developed to help explain and expand upon the Buddha's teachings. For others still, they see the dharma as referring to the "truth" or ultimate reality or "the way things are".

The Dharma is one of the Three Jewels of Buddhism of which practitioners of Buddhism seek refuge in (what one relies on for his/her lasting happiness). The three jewels of Buddhism are the Buddha (mind's perfection of enlightenment), the Dharma (teachings and methods), and the Sangha (awakened beings who provide guidance and support).

Buddha's Dharma Body: The qualities of the Dharma (Law, truth) is the same as the qualities of the Buddha and forms his "truth body" or "Dhamma Kaya": In the Samyutta Nikaya, Vakkali Sutta, Buddha said to his disciple Vakkali that,

"Yo kho Vakkali dhammam passati so mam passati"

Vakkali, whoever sees Dhamma, sees me [the Buddha]

Another reference from the Agganna Sutta of the Digha Nikaya, says to his disciple Vasettha:

"Tathagatassa h'etam Vasettha adivacanam Dhamma-kayo iti pi...":

O Vasettha! The Word of Dhammakaya is indeed the name of the Tathagata

Knowing these attributes, Buddhists believe that they will attain the greatest peace and happiness through the practice of the Dhamma. Each person is therefore fully responsible for himself to put it in the real practice.

Here the Buddha is compared to an experienced and skilful doctor, and the Dhamma to proper medicine. However efficient the doctor or wonderful the medicine may be, the patients cannot be cured unless they take the medicine properly. So the practice of the Dhamma is the only way to attain the final deliverance of Nibbana.

These teachings ranged from understanding karma (Pali: kamma) (cause and effect) and developing good impressions in one's mind, to how to reach full enlightenment by recognizing the nature of mind.

Dharmas in Buddhist Phenomenology: Other uses include dharma, normally spelled with a small "d" (to differentiate), which refers to a phenomenon or constituent factor of human experience. This was gradually expanded into a classification of constituents of the entire material and mental world. Rejecting the substantial existence of permanent entities which are qualified by possibly changing qualities, Buddhist Abhidharma philosophy, which enumerated seventy-five dharmas, came to propound that these "constituent factors" are the only type of entity that truly exists.

This notion is of particular importance for the analysis of human experience: Rather than assuming that mental states inhere in a cognizing subject, or a soul-substance, Buddhist philosophers largely propose that mental states alone exist as "momentary elements of consciousness", and that a subjective perceiver is assumed.

One of the central tenets of Buddhism, is the denial of a separate permanent "I", and is outlined in the three marks of e. The three signs: 1. Dukkha (Pali: Dukkha)-Suffering, 2. Anitya (Pali: Anicca)-Change/Impermanence, 3. Antman (Pali: Anatta)-Non-self. At the heart of Buddhism, is the denial of a "self" or "I" (and hence the delusion) as a separate self-existing entity.

Later, Buddhist philosophers like Nagarjuna would question whether the dharmas (momentary elements of consciousness) truly have a separate existence of their own. (*i.e.* Do they exist apart from anything else?) Rejecting any inherent reality to the dharmas, he asked (rhetorically):

What is impermanent and permanent? What is neither?

Auspicious is the pacification of phenomenal metastasis, the pacification of all apprehending;

There is no dharma whatsoever taught by the Buddha to whomever, whenever, wherever. Dharma as Righteousness.

According to S.N. Goenka, teacher of Vipassana Meditation, the original meaning of dhamma is "dhareti ti dhamma', or "that which is contained". Dharma in the Buddhist scriptures has a variety of meanings, including "phenomenon", and "nature" or "characteristic". Dharma also means 'mental contents', and is paired with citta, which means heart/mind. The pairing is paralleled with the pairing of kaya (body) and vedana (feelings or sensations, that which arise within the body but experienced through the mind), in major sutras such as the Mahasatipatthana sutra. Dharma is also used to refer to the teachings of the Buddha, not in the context of the words of one man, even an enlightened man, but as a reflection of natural law which was re-discovered by this man and shared with the world. A person who lives their life with an understanding of this natural law, is a "dhammic" person, which is often translated as "righteous".

Dharma (Sanskrit) or Dhamma (Pali) in Buddhism has two primary meanings: the teachings of the Buddha which lead to enlightenment, and the constituent factors of the experienced world. In East Asia, the character for Dharma is Ol, pronounced fI in Mandarin and hM in Japanese. The Tibetan translation of this term is chos (cho).

Samsara (Buddhism): Samsara, the Sanskrit and Pali term for "continous movement" or "continuous flowing" refers in Buddhism to the concept of a cycle of birth (jati) and consequent decay and death (jarmarana), in which all beings in the universe participate and which can only be escaped through enlightenment. It is generally opposed to Nirvana or nibbna.

Samsara in Nikaya Buddhism: Whereas in Hinduism some being (atman, jiva, etc.) is regarded as being subject to Samsara, Buddhism was founded on a rejection of such metaphysical substances, and originally accounts for the process of rebirth/reincarnation by appeal to phenomenological or psychological constituents. Later schools of Buddhism such as the Pudgalavda, however, re-introduce the concept of a "person" which transmigrates. The basic idea that there is a cycle of birth and rebirth is, however, not questioned in early Buddhism and its successors, and neither is, generally, the concept that samsara is a negative condition to be abated through religious practice concluding in the achievement of final Nirvana.

Samsara in Mahayana Buddhism: According to several strands of the Mahayana Buddhist tradition, the division of samsara and Nirvana is attacked using an argument that extends some of the basic premises of antman and of Buddha's attack on orthodox accounts of existence. This is found poetically in the "Perfection of Wisdom" literature and more analytically in the philosophy of Nagarjuna and later writers. It is not entirely clear which aspects of this theoretical move were developed first in the sutras and which in the philosophical tradition.

Samsara in Tibetan Buddhism: Samsara is uncontrollably recurring rebirth, filled with suffering and problems (according to Kalacakra tantra as explained by Dr. A. Berzin).

Rebirth (Buddhism)

Rebirth in Buddhism is the doctrine that streams of consciousness composed of aggregates called skandha causes a new form to emerge and would continue as long as those aggregates abide in ignorance or avidya (Buddhism), which is the root cause of existence in Buddhism.

Although the cessation of a life is not in itself a sufficient condition for the inception of a new life (since arhats, pratyekabuddhas and buddhas pass away without rebirth), the supporting conditions for a new birth are almost always present. From an external perspective, each life appears as a link in a beginningless sequence of lives, varying in length and in quality.

In traditional Buddhist cosmology, these lives can be in any of a large number of states of being, including those of humans, any kind of animal, and several types of supernatural being (see Six realms). The type of rebirth that arises at the end of one life is conditioned by the karmas (actions of body, speech and mind) of the previous life; good karmas will yield a happier rebirth, bad karmas will produce one which is more unhappy.

In the traditional Buddhist languages of Sanskrit and Pali, there is no word corresponding exactly to the English "rebirth".

A rebirth, that is, the state one is born into, is referred to as jati, *i.e.* simply "birth", also referring to the process of being born or coming into the world in any way. The entire process of change from one life to the next is called punarbhava (Sanskrit) or punabbhava (Pali), literally "becoming again"; it is also known simply as bhava, *i.e.* "becoming". The process seen from a universal perspective, encompassing all living beings, is called samsara.

From an interior perspective, a person who remembers or imagines a past life is likely to think of it as representing a continuity of existence between lifespans, *i.e.*, that the same person (however defined) was formerly one person (with a certain name and body) and is now a different person (with another name and body). This perspective is objectionable from the point of view of Buddhist philosophy on two counts. First, because it seems to postulate an enduring, self-existing entity that exists separate from the elements of mind and body, contrary to the Buddhist philosophical position of antman. Second, because it overlooks the characterization of this process as one of constant change, both within and between lives, in which the newly-arising life is conditioned by but in no respect identical to the predecedent life.

Nonetheless, the Buddha is represented using language reflecting the interior perspective in stories about his past lives in both jatakas and sutras. For instance, "At that time I was the Brahmin, the Great Steward..." (Mahgovinda-sutta, DN.19) or "Six times, nanda, I recall discarding the body in this place, and at the seventh time I discarded it as a wheel-turning monarch..." (Mahsudassana-sutta, DN.17). This can be regarded as a concession to the needs of conventional speech.

Rebirth as Cycle of Consciousness: Another view of rebirth describes the cycle of death and rebirth in the context of consciousness rather than the birth and death of the body. In this view, remaining impure aggregates, skandhas, reform consciousness into a new form.

Buddhist meditation teachers suggest that through careful observation of the mind, it is possible to see consciousness as being a sequence of conscious moments rather a continuum of awareness. Each moment is an experience of an individual mind-state: a thought, a memory, a feeling, a perception. A mind-state arises, exists and, being impermanent, ceases following which the next mind-state arises. Thus the consciousness of a sentient being can be seen as a continuous series of birth and death of these mind-states. In this context rebirth is simply the persistence of this process. Clearly this explanation of rebirth is wholly divorced from rebirth which may follow bodily death and it is possible for a Buddhist to believe in either, both or neither definition.

The explanation of rebirth as a cycle of consciousness is consistent with other core Buddhist beliefs, such as anicca (impermanence), dukkha (unsatisfactoriness) and anatta (non-self). Furthermore it is possible to observe a karmic link between these mind-states.

In the practice of Vipassana meditation, the meditator uses "bare attention" to observe the endless round of mind-states. This observation derives insight and understanding from seeing this cycle of birth, death and rebirth without interfering, owning or judging the individual states of mind that arise and pass away. This understanding enables them to limit the power of desire, which according to the second noble truth of Buddhism is the cause of Dukkha (suffering, unsatisfactoriness) thus making possible the realisation of Nibbana. So it can be concluded that the understanding of rebirth in the context of the cycle of consciousness is an invaluable and practical component of the fundamental aim of Buddhism.

Rebirth as Buddhist Reincarnation: Within Buddhism, the term rebirth or re-becoming (Sanskrit: punarbhava) is preferred to "reincarnation", as the latter is taken to imply there is a fixed entity that is reborn. However, this still leaves the question as to what exactly the process of rebirth entails.

The lack of a fixed self does not mean lack of continuity. One of the metaphors used to illustrate this is that of fire. For example, a flame is transferred from one candle to another, or a fire spreads from one field to another. In the same way that it depends on the original fire, there is a conditioned relationship between one life and the next; they are not identical but neither are they completely distinct. The early Buddhist texts make it clear that there is no permanent consciousness that moves from life to life.

Early Buddhists had to deal with the problems of establishing the nature of the causal link between two lives, especially the crucial one of how one being could receive the fruits of the actions of a previous being, now dead, and how saCskras, or volitional tendencies to act and think in particular ways can be transferred from one being to another.

The Puggalavda school (now extinct) believed in a personal entity (puggala) separate from the five skandhas that provided a link of personal continuity that allows for karma to act on an individual over time. The medieval Pali scholar Buddhaghosa posited a 'rebirth-linking consciousness' (patisandhi), which connected the arising of a new life with the moment of death, but how one life came to be associated with another was still not made clear.

Some schools were led to the conclusion that karma continued to exist in some sense and adhere to a particular person until it had worked out its consequences. Another school, the Sautrantika, made use of a more poetic model to account for the process of karmic continuity. For them, each act 'perfumed' the individual and led to the planting of a 'seed' that would later germinate as a good or bad karmic result.

While all Buddhist traditions seem to accept some notion of rebirth, there is no unified view about precisely how events unfold after the moment of death. Theravada Buddhism generally asserts that rebirth is immediate. The Tibetan schools, on the other hand, hold to the notion of a bardo (intermediate state) which can last up to forty-nine days, and this has led to the development of a unique 'science' of death and rebirth, a good deal of which is set down in what is popularly known as The Tibetan Book of the Dead. Also, Rick Strassman's book "The Spirit Molecule" touches on the association between the intermediate state and a possible scientific explanation.

While Theravada Buddhism generally denies there is an intermediate state, some early Buddhist texts seem to support it. One school that adopted this view was the Sarvstivda, who believed that between death and rebirth there is a sort of limbo in which beings do not yet reap the consequences of their previous actions but in which they may still influence their rebirth. The death process and this intermediate state were believed to offer a uniquely favourable opportunity for spiritual awakening.

There are many references to rebirth in the early Buddhist scriptures. These are some of the more important: Mahakammavibhanga Sutta (Majjhima Nikaya 136); Upali Sutta (Majjhima Nikaya 56); Kukkuravatika Sutta (Majjhima Nikaya 57); Moliyasivaka Sutta (Samyutta Nikaya 36.21); Sankha Sutta.

Rebirth in the Context of Other Religions and Other Buddhist Beliefs: In the religions of Middle Eastern origin, Judaism, Christianity and Islam, life and death are believed to be linear: a being is born (usually understood as a new creation), lives, and then dies, at which point their soul or other part that survives death, passes to a domain that is inaccessible to living beings and remains there indefinitely, or until the end of the world. (Note that reincarnation, in the limited form of gilgul neshamot plays a role in some forms of Judaism. An even more restricted belief in reincarnation (tanasukh) is found in the Druze religion which is derived from Islam.)

The Buddha lived at a time of great philosophical creativity in India, and many different concepts of the nature of life and death that were proposed at that time. Some thinkers were materialists, believing that there was no existent consequent upon the end of a life, and that there was an atman (self) which was annihilated upon death. Others believed in a form of cyclic existence, where a being is born, lives, dies and then is re-born, but in the context of a type of determinism or fatalism, in which karma played no role. Others were "eternalists", postulating an eternally existent atman, comparable to the Western concept of the soul: when a being (or his body) dies, the atman survives death and is re-embodied (reincarnates) as another living being, based on its karmic inheritance. This last belief is the one that has come to be dominant (with certain modifications) in modern Hinduism.

The Buddha taught a concept of rebirth that was distinct from that of any Indian teacher contemporary with him. This concept was consistent with the common notion of a sequence of related lives stretching over a very long time, but was constrained by two core Buddhist concepts: anatt, that there is no irreducible atman or "self" tying these lives together; and anicca, that all compounded things are subject to dissolution, including all the components of the human person and personality. The Buddha's detailed conception of the connections between action (karma), rebirth, and their ultimate causes is set out in the twelve links of dependent origination.

Shunyata

Zknyat, (Sanskrit) or Sunnat (Pali) is a term, translated as "Emptiness" or "Voidness", which constitutes an aspect of the Buddhist metaphysical critique as well as Buddhist epistemology and phenomenology. Zknyat signifies that everything one encounters in life is empty of soul, permanence, and self-nature. Everything is inter-related, never self-sufficient or independent; nothing has independent reality. Yet [sunyat never connotes nihilism, which Buddhist doctrine considers to be a delusion, just as it considers materialism to be a delusion.

In the Mahayana Tathagatagarbha sutras, in contrast, only impermanent, changeful things and states (the realm of samsara) are said to be empty in a negative sense-but not the Buddha or Nirvana, which are stated to be Real, eternal and filled with inconceivable, enduring virtues.

Origin and Development of Zknyat: The theme of Sunyat emerged from the Buddhist doctrines of Anatta (Pali, Sanskrit:Antman—the nonexistence of the self, or atman) and Paticcasamuppada (Pali, Sanskrit: pratityasamutpda, Interdependent Arising). The Sunna Sutta, part of the Pali Canon, relates that the monk Ananda, the attendant to Gautama Buddha asked, "It is said that the world is empty, the world is empty, lord. In what respect is it said that the world is empty?" The Buddha replied, "Insofar as it is empty of a self or of anything pertaining to a self: Thus it is said, Ananda, that the world is empty." Zknyat is a key theme of the Heart Sutra, which is commonly chanted by Mahayana Buddhists worldwide. Zknyat was further developed by Nagarjuna and the Madhyamaka school, which is usually counted as an early Mahayana school. Zknyat ("positively" interpreted-see Tathagatagarbha section below) is also an important element of the Tathagatagarbha literature, which played a formative role in the evolution of subsequent Mahayana doctrine and practice. It should be noted that the exact definition and extent of shunyata varies within the different Buddhist schools of philosophy. In the Tibetan Buddhist tradition, detailed dialogs between the perspectives of the various schools are preserved in order to train students.

Zknyat in Presecular Buddhism, in the Nikayas: Sunnata "Emptiness" (sunnata) in Pali contexts is not the metaphysical Zero (Nonbeing as the principle of Being, Infinite Possibility as distinguished from Indefinite Actuality), but a characteristic of this world, where it has been explained that when the Almsman returns from a deathlike Contemplation in which consciousness and feeling have been arrested, "three touches touch him," "emptiness (sunnato)," "formlessness (animito)" and "making no plans (appanihito phasso)," and he discriminates (viveka) accordingly; and the meaning of "emptiness" 'is explained at M 1.29, "emancipation of the mind by Emptiness (sunnata ceto vimutti) being consequent upon the realization that this world is empty of spirit or anything spiritual'; sunnata is synonymous with anatta; of which it really only paraphrases and isolates the privative AN.

It is no doubt in the same sense that in A 1.72, "the texts are coupled with `emptiness' (suttanta... sunnata patisannuta)"; there is, in fact, nothing more characteristic of Buddhist teaching than its constant resort to negatives (above all in the sense of the word anatta), which even some contemporary hearers found perplexing. The denial of spirituality to contingent things in particular is a denial of any real essence to these things in themselves, and thus forms the basis of the more sweeping sunyavada doctrine which in the Mahayana denies not any "value" but any essence to even the Buddha's appearance and to the promulgation of the Dhamma itself.

If such a doctrine disturbs us, it may be found more palatably expressed in the Vajracchedika Sutra thus, "Those who see me in the body (rupena) and think of me in sounds (ghosaih), their way of thinking is false, they do not see me at all.... The Buddha cannot be rightly understood (rjuboddhum) by any means (upayena). Not that "means" are not dispositive to a right understanding, but that if regarded as ends, even the most adequate means are a hindrance. In such a radical iconoclasm as this all traditional teachings are finally agreed. What is true of ethics is also true of the supports of contemplation: as in the well known Parable of the Raft, the means are of no more use when the goal has been reached.

"What is non-emptiness' monks? It is the Soul (attano), Self-Nature (sabhava) is non-emptiness, is the refuge to be sought". What friend is emancipation of the mind by means of devoidness (sunnata)? Herein a follower has gone to a clearing in the forest and the root of a tree and investigates thusly: 'This is devoid (sunnamidam) of the Soul and what the Soul subsists upon." This is called emancipation of the mind by means of devoidness (sunnata cetovimmuti). The fixed unshakable emancipation of the mind is devoid of lusts, devoid of hate, and devoid of delusions. "What is shunyata liberation? He considers: 'this is void of the Soul and what belongs to the Soul'.

Zunyata in the Heart Sutra: The Heart Sutra (one of the Mahayana Perfection of Wisdom Sutras) declares that the skandhas, which constitute our mental and physical existence, are empty in their nature or essence, *i.e.*, empty of any such nature or essence. But it also declares that this emptiness is the same as form (which connotes fullness)—*i.e.*, that this is an emptiness which is at the same time not different from the kind of reality which we normally subscribe to events; it is not a nihilistic emptiness that undermines our world, but a "positive" emptiness which defines it.

- "The noble bodhisattva, Avalokitesvara, engaged in the depths of the practice of the perfection of wisdom, looked down from above upon the five skandhas (aggregates), and saw that they were empty in their essential nature."

- "Here, O Sariputra, emptiness is form; form is just emptiness. Apart from form, emptiness is not; emptiness, form is not. Emptiness is that which is form, form is that which is emptiness. Just thus are perception, cognition, mental construction, and consciousness."
- "Here, O Sariputra, all phenomena of existence are marked by emptiness: not arisen, not destroyed, not unclean, not clean not deficient nor fulfilled."

Zknyat in Nagarjuna: For Nagarjuna, who provided the most important philosophical formulation of [sunyat, emptiness as the mark of all phenomena means is a natural consequence of dependent origination; indeed, he identifies the two. In his analysis, any enduring essential nature (*i.e.*, fullness) would prevent the process of dependent origination, would prevent any kind of origination at all, for things would simply always have been and always continue to be. That things happen is proof that things lack the kind of nature attributed to them in mainstream Indian metaphysics.

An interesting consequence of this is that this enables Nagarjuna to put forth a bold argument regarding the relation of Nirvana and samsara. If all phenomenal events (*i.e.*, the events that constitute samsara) are empty, then they are empty of any compelling ability to cause suffering. For Nagarjuna, Nirvana is neither something added to samsara nor any process of taking away from it (*i.e.*, removing the enlightened being from it). In other words, Nirvana is simply samsara rightly experienced in light of a proper understanding of the emptiness of all things.

Sunyata in the Tathagatagarbha Sutras: The class of Buddhist scriptures known as the Tathagatagarbha sutras presents a seemingly variant understanding of Emptiness, according to which the Buddha and Nirvana, unlike compounded, conditioned phenomena, are not empty of intrinsic existence, but merely empty of the impermanent, the painful and the Self-less. In the "Srimala Sutra" the Buddha is seen as empty of all defilement and ignorance, not of intrinsic Reality. The "Mahayana Mahaparinirvana Sutra" supports such a vision and views Ultimate Emptiness as the Buddhic cognition ("jnana") which perceives both Emptiness and non-Emptiness, wherein "the Empty is the totality of Samsara and the non-Empty is Great Nirvana".

The Buddha in the Mahayana Mahaparinirvana Sutra further indicates that to view absolutely everything as empty is an unbalanced approach and constitutes a deviation from the middle path of Buddhism: "The wise perceive Emptiness and non-Emptiness, the Eternal and the Impermanent, Suffering and Bliss, the Self and the non-Self.... To perceive the Emptiness of everything and not to perceive non-Emptiness is not termed the Middle Way; to perceive the non-Self of everything and not to perceive the Self is not termed the Middle Way."

Moreover, this particular sutra contains a passage in which the Buddha castigates those who view the Tathagatagarbha (which is the indwelling, immortal Buddha-element) in each being as empty, declaring of them that they are effectively committing a form of painful spiritual suicide through their wrongheaded stance: "By having cultivated non-Self in connection with the Tathagatagarbha and having continually cultivated Emptiness, suffering will not be eradicated but one will become like a moth in the flame of a lamp." (Tibetan version of the Mahayana Mahaparinirvana Sutra). The attainment of nirvanic Liberation ("moksha"), by contrast, is said to open up a realm of "utter bliss, joy, permanence, stability, [and] eternity" (ibid), in which the Buddha is "fully peaceful" (Dharmakshema "Southern" version). Perhaps the clearest statement of Tathagatagarbha Buddhism's understanding of Emptiness is found in the Angulimaliya Sutra, where we read the following clarifying explanation:

> *"... by cultivating extreme emptiness and continually considering things to be empty, one will behold the utter destruction of all phenomena. Though Liberation is not empty, one will see and think it to be empty. Thus, for example, having thought hail-stones to be jewels, one comes to think that real gems are empty [[knya]. Likewise, you too think of phenomena which are not empty [a[knya] to be empty [[knya], for viewing phenomena as empty,*

you dissolve into emptiness ([knya) even those phenomena which are not empty. Some phenomena are empty [of existence] and some phenomena are not empty [of existence]. Just like the hail-stones, the billions of kleshas [mental and moral afflictions] are empty [of existence], like the hail-stones, those phenomena appertaining to ignorance are empty [of existence] and swiftly fade away. Like the real beryl gems, the Buddha is eternal. Liberation is like the real beryl gems."

Thus in the distinctive Tathagatagarbha sutras a balance is drawn between the empty, impermanent and coreless realm of samsara and the lasting, liberating Reality of the Buddha and Nirvana.

Pratitya-samutpada: The doctrine of Pratityasamutpda (Sanskrit) or Pamiccasamuppda. Dependant Arising is an important part of Buddhist metaphysics. Common to all Schools of Buddhism, it states that phenomena arise together in a mutually interdependent web of cause and effect. It is variously rendered into English as "dependent origination," "conditioned genesis," "dependent co-arising," "interdependent arising," etc.

Dependent Origination

The enlightenment (Bodhi) of the Buddha Gautama was simultaneously his liberation from suffering and his insight into the nature of the universe – particularly the nature of the lives of 'sentient beings' (principally humans and animals). What the Buddha awakened to (Bodhi means 'to awaken') was the truth of dependent origination.

This is the understanding that any phenomenon 'exists' only because of the 'existence' of other phenomena in an incredibly complex web of cause and effect covering time past, time present and time future. Because all things are thus conditioned and transient (anicca), they have no real independent identity (anatta) so do not truly 'exist', though to ordinary deluded minds this appears to be the case. All phenomena are thus fundamentally insubstantial and 'empty' (sunya). Wise human beings, who 'see things as they are' (yatha-bhuta-nana-dassana), renounce attachment and clinging, transform the energy of desire into awareness and understanding, and eventually transcend the conditioned realm of form becoming Buddhas or Arahants.

General Formulation

A general formulation of this concept goes:

With this as condition,

That arises.

With this NOT as condition,

That does NOT arise.

An example to illustrate:

You go on summer holiday to a hot climate, such as Arizona, Spain or Australia. It's a hot clear day and you're sunbathing by the hotel pool with the sun beating down on you. You will begin to feel hot, sweaty, uncomfortable, and soon feel thirsty. You go get yourself a drink to quench your thirst, and think "It's too hot to sit by the pool today, I'm going back to my hotel room where it's cooler, to read for a while".

With "hot summer sun" as condition,

Sweat, thirst and discomfort arises.

With "cool hotel room" as condition,

Sweat, thirst and discomfort do NOT arise.

This draws attention to the constant flux of "Coming to be, and Ceasing to be" that is happening all the time. All phenomena are subject to this unending interaction. And since all phenomena are dependent on other phenomena, they are all transient and impermanent.

Applications

The general formulation has two very well known applications.

Four Noble Truths

The first application is to suffering, and is known as the Four Noble Truths:

1. *Dukkha:* There is suffering. Suffering is an intrinsic part of life also experienced as dissatisfaction, discontent, unhappiness, impermanence.
2. *Samudaya:* There is a cause of suffering, which is attachment or desire (tanha).
3. *Nirodha:* There is a way out of suffering, which is to eliminate attachment and desire.
4. *Marga:* The path that leads out of suffering is called the Noble Eightfold Path.

Twelve Nidanas

The other application is to the rebirth process and is known as the Twelve Nidanas or the Twelve Links of Conditioned Existence. In this application of pratitya-samutpada, each link is conditioned by the preceding one, and itself conditions the succeeding one. These cover three lives:

Former Life

- ignorance
- activities which produce karma

Current Life

- consciousness
- name and form (personality or identity)
- the twelve domains (5 physical senses + the mind + forms, sounds,..., thoughts)
- contact (between objects and the senses)
- sensation (registering the contact)
- desire (for continued contact)
- attachment

Future Life

- becoming (conception of a new life)
- birth
- old age and death.

With respect to the destinies of human beings and animals, dependent origination has a more specific meaning as it describes the process by which such sentient beings incarnate into any given realm and pursue their various worldly projects and activities with all the concomitant suffering involved. Among these sufferings are ageing and death.

Ageing and death are experienced by us because birth and youth have been experienced. Without birth there is no death. One conditions the other in a mutually dependent relationship. Our becoming in the world, the process of what we call 'life', is conditioned by the attachment and clinging to certain ideas and projects such as having a family or making money. This attachment and clinging in turn cannot exist without craving as its condition.

The Buddha understood that craving comes into being because there is sensation in the body which we experience as pleasant, unpleasant or neutral. When we crave certain things such as alcohol, sex or sweet foods it is the sensation induced by contact with the desired object that we crave rather than the object itself. Sensation is caused by contact with such objects of the senses. The contact or impression made upon the senses (manifesting as sensation) is itself dependent upon the six sense organs which themselves are dependent upon a psycho-physical entity such that a human being is. The whole process is summarised by the Buddha as follows:

> *The thrust of the formula is such that when certain conditions are present, they give rise to subsequent conditions, which in turn give rise to other conditions and the cyclical nature of life in Samsara can be seen. This is graphically illustrated in the Bhavacakra (Wheel of life).*

There appears to be widespread misunderstanding of the formula in relation to time scales. Many references made to Pratityasamutpda are expressed over lifetimes. While this is true in the wider sense, more practically, this is to be seen as a daily cycle occurring from moment-to-moment throughout each day.

Its necessary to refer to the above in order to fully understand and make use of Pratitya-samutpada.

For example, Avidy the first condition, its necessary to refer to Three Signs of Being for a fuller explanation and a better understanding of its fit and function within Pratitya-samutpada. Its also necessary to understand the Three Fires and how they fit in the scheme of things.

A quick glance at the Bhavacakra, you will note that the Three Fires sit at the very center of the schemata, and drive the whole edifice.To understand this, it is necessary to study different sorts of conditions, because only one of them is called "causal condition". Example, ignorance can determine activities as an "object condition" if one... decide to reduce ignorance. And then because in this life one has been ignorant, and acted in such a way as to produce karma, the cycle continues round again.

Nibbana (Skt Nirvana) is often conceived of as stopping this cycle. By removing the causes for craving, craving ceases. So with the ceasing of birth, death ceases. With the ceasing of becoming, birth ceases... and so on until with the ceasing of ignorance no karma is produced, and the whole process of death and rebirth ceases. In fact the opportunity for change comes between the stages of sensation and desire, since as we saw above it is craving that drives the whole process. If one can simply experience sensations without desiring, then craving will not arise, and one can begin to be free from the cycle of birth and death.

Madhyamaka and Pratitya-samutpada

Though the formulations above appear might seem to imply that pratitya-samutpada is a straightforward causal model, in the hands of the Madhyamaka school, Pratitya-samutpada is used to demonstrate the very lack of inherent causality, in a manner that appears somewhat similar to the ideas of David Hume.

The conclusion of the Madhyamikas is that causation, like being, must be regarded as a merely conventional truth and that to take it as really (or essentially) existing would be both a logical error and a perceptual one, arising from ignorance and a lack of spiritual insight.

According to the analysis of Nagarjuna, the most prominent Madhyamika, true causality depends upon the intrinsic existence of the elements of the causal process (causes and effects), which would violate the principle of anatta, but pratitya-samutpda does not imply that the apparent participants in arising are essentially real.

Because of the interdependence of causes and effects (*i.e.* causes depend on their effects in order to be causes, and effects likewise depend on their causes in order to be effects), it is quite meaningless to talk about them as existing separately. However, the strict identity of cause and effect is also refuted, since if the effect were the cause, the process of origination could not have occurred. Thus both monistic and a dualistic accounts of causation are rejected. Therefore Nagarjuna explains that the anatta (or emptiness) of causality is demonstrated by the interdependence of cause and effect, and likewise that the interdependence (pratitya-samutpda) of causality itself is demonstrated by its anatta. In his Entry to the middle way, Candrakirti asserts, "If a cause produces its requisite effect, then, on that very account, it is a cause. If no effect is produced, then, in the absence of that, the cause does not exist."

Pratitya-samutpada in Dzogchen: In Dzogchen tradition the interdependent origination is considered illusory: '(One says), "all these (configurations of events and meanings) come about and disappear according to dependent origination." But, like a burnt seed, since a nonexistent (result) does not come about from a nonexistent (cause), cause and effect do not exist.

Being obsessed with entities, one's experiencing itself [sems, citta], which discriminates each cause and effect, appears as if it were cause and condition.' (from byang chub sems bsgom pa by Manjusrimitra. Primordial experience. An Introduction to Dr. Zogs-chen Meditiation, pp. 60, 61)

The Reversibility of Dependent Arising: Pratitya-samutpada is most commonly used to explain how suffering arises depending on certain conditions, the implication being that if one or more of the conditions are removed (if the "chain" is broken), suffering will cease.

There is also a text, the Upanisa Sutta in the Samyutta Nikaya, in which a discussion of the conditions not for suffering but for enlightenment are given. This is sometimes glossed as "transcendental" dependent arising. The chain in this case is:

- Suffering [of the rounds of rebirth]
- Faith
- Gladness
- Rapture
- Tranquillity
- Happiness
- Concentration
- Knowledge and vision of things as they really are
- Revulsion
- Dispassion
- Liberation
- Knowledge of destruction of the poisons

Karma in Buddhism

Karma (Sanskrit karman) or Kamma (Pali) means "action" or "doing"; whatever one does, says, or thinks is a karma.

In Buddhism, the term karma is used specifically for those actions which spring from :

- mental intent
- mental obsessions

Which bring about a fruit or result (vipka), either within the present life, or in the context of a future rebirth. Karma is the engine which drives the wheel of samsara for each being. In a discourse (Anguttara Nikaya Nibbedhika Sutta) the Buddha said :

> *"Intention, monks, is kamma I say. Having willed, one acts through body, speech and mind".*

Karma is used as an ethical principle, rather than a cosmological explanation for the world. Buddhists believe that the actions of beings will affect their own future, and because of this there are no private actions: all actions have a consequence.

The emphasis of karma in Buddhism is on cause, not on effect: Buddhists do not say "it was due to her karma that it happened to her"-indeed the karmic consequences of one's actions are dependent on sufficient conditions.

Buddhism distinguishes samsaric happiness (birth in the high realms), from the final state of enlightenment: nirvana; so likewise there is samsaric good karma, which leads to the high realms (such as the human realm), and then there is liberating karma-which is supremely good.

Therefore the major dichotomy is samsaric karma and liberating karma, of which the former is typically divided into the three: good, neutral, and bad (in accordance with the degree of samsaric happiness or suffering that will mature as a consequence). Likewise, liberating karma is divided into three: according to whether it will mature the being as a [rvakabuddha, pratyekabuddha, or a samyaksambuddha.

In Buddhism, the term karma is often used to refer only to samsaric karma, as indicated by the twelve nidanas of dependent origination.

The differentiation between "good" karma and "liberating" karma has been used by some scholars to argue that the development of Tantra depended upon Buddhist ideas and philosophies.

Because of the inevitability of consequence, Karma entails the notion of Buddhist rebirth. However, karma is not the sole basis of rebirth. The rebirths of eighth stage (and above) Bodhisattvas in the Mahayana tradition refers to those liberated beings who consciously choose to be reborn in a future life in order to help others still trapped in samsara.

Some people have problems with the teaching on karma, often of what exactly the Buddha is asking them to believe in when asking them to have conviction in karma.

- First, action really is happening — it's not an illusion.
- Second, you really are responsible for your actions. There's no outside force like the stars or some good or evil being acting through you. When you're conscious, you're the one who decides what to do.

- Third, your actions have results — you're not just writing on the water — and those results can be good or bad depending on the quality of the intention behind the act.

Wrong Understandings of Karma: In Buddhism, Karma is not pre-determinism, fatalism or accidentalism, as all these ideas lead to inaction and destroy motivation and human effort that a human being can change for the better no matter what his or her past was and are designated as "wrong view" in Buddhism.

1. Pubbekatahetuvada: The belief that all happiness and suffering arise from previous kamma (Past-action determinism).
2. Issaranimmanahetuvada: The belief that all happiness and suffering are caused by the directives of a Supreme Being (Theistic determinism).
3. Ahetu-apaccayavada: The belief that all happiness and suffering are random, having no cause (Indeterminism or Accidentalism).

"Bhikkhus, adhering to previously done kamma as the essence, there are neither motivation nor effort with what should be done and what should not be done... Not upholding ardently what should be done, nor abandoning what should be abandoned, those ascetics and Brahmins are as if deluded, lacking a control, incapable of having any true teaching. (Buddha). In Buddhism, Karma is simply there as a guide and an indication of what your present is and how one's future can be made better by self effort, otherwise, fatalism and pre-determinism is the anti-thesis of the notion of perfection or self-conquest-which is the primary aim of Buddhism.

Buddhist Meditation: Buddhist meditation, meditation used in the practice of Buddhism, "includes any method of meditation that has Enlightenment as its ultimate aim". The closest word for meditation in the classical languages of Buddhism is bhavana or "mental development".

Methods of Meditation: The main methods of Buddhist meditation are divided into samatha (tranquility meditations) and vipassana (insight meditations). The samatha meditations include anapana (mindfulness of breathing) and mettabhavana (development of loving kindness). The vipassana meditations includes contemplation on impermanence, the six element practice, and contemplation on conditionality. Samatha meditations usually precede and prepare for vipassana meditations.

Each of the five basic methods is an "antidote" to one of the five mental "poisons":

Meditation type	*Method*	*Counteracts*	*Develops*
Samatha (tranquility meditations)	anapana	distraction	concentration
	metta bhavana	hatred [and sentimental attachment]	love
Vipassana (insight meditations)	contemplation of impermanence	craving	inner peace, freedom
	six element practice	conceit	clarity regarding nature of self
	contemplation of conditionality	ignorance	wisdom, compassion
Samatha (tranquility meditations)	karuna bhavana	cruelty, sentimental pity and horrified anxiety	compassion
	mudita bhavana	resentment, envy and vicarious enjoyment	sympathetic joy
	upekkha bhavana	fixed indifference and apathetic neutrality	equanimity

Not included under samatha in the above table are all the four brahma-viharas (lit. "sublime abodes") of which mettabhavana is the most often practiced one:

Anapanasati: Anapanasati, (Pali) meaning mindfulness of breathing ("sati" means mindfulness, "napana" refers to the breath moving in and out) is a basic form of meditation taught by the Buddha. According to the Buddha's teaching in the Anapanasati Sutta, practicing mindfulness of breathing meditation as a part of the Noble Eightfold Path leads to the removal of all defilements (kilesa) and finally the attainment of Nibbana (Nirvana). The Buddha's teaching was based on his own experience in using anapanasati as part of his means of achieving his own enlightenment. Mindfulness of breath is often practiced along with mettabhavana to prevent withdrawal from the world and loss of compassion (Kamalashila 1996).

The Anapanasati Sutta is specifically about the mindfulness of in-and-out breathing. It recommends the practice of napanasati meditation as a means of cultivating the seven factors of awakening: sati (mindfulness), dhamma vicaya (analysis), viriya (persistence), which leads to piti (rapture), then to passaddhi (serenity), which in turn leads to samadhi (concentration) and then to upekkha (equanimity). Finally, the Buddha taught that with these factors developed in this progression, that the practice of napanasati would lead to release or Nibbana (Nirvana).

Anapanasati meditation is normally practiced either in conjunction with vipassana meditation, with zazen, or zen meditation in the soto Zen tradition (shikantaza). However, before the meditator can use vipassana, the meditator must first develop enough concentration using anapanasati meditation to be able to undertake vipassana meditation.

Practice: The practice of anapanasati varies. Typically, to begin, one needs to sit in a comfortable position, with the back and neck straight. It is also helpful if one can find a comfortable and peaceful environment.

Breathe naturally, without attempting to change the length or the depth of the breath. If the breath is short, then observe that the breath is short. If the breath is long, then observe that the breath is long.

While breathing, the meditator trains the mind to be sensitive to or focused on various things, and calming or steadying various things. The meditator trains in breathing in sensitive to one or more of: the entire body, rapture, pleasure, the mind, or mental processes, and then breathing out sensitive to this thing. The meditator trains in breathing in focusing on one or more of inconstancy, dispassion, cessation and relinquishment, and then breathing out focusing on this thing. The meditator trains in breathing in steadying, satisfying or releasing the mind, and then breathing out steadying, satisfying or releasing the mind.

Tutors will explain that an untrained mind constantly has thoughts interrupting one's focus. Like the waves in the ocean, thoughts arise, and they slowly fall away. If one disregards the thoughts, they slowly wither and disappear. On the other hand, if one pays attention, then soon one is lost in a web of thoughts.

In this tradition there are two types of thoughts: thoughts from the past and thoughts about the future. It is said that when left unattended, the mind will flit from one thought to another, wandering aimlessly. There are thoughts which bring happiness, and there are thoughts which bring sadness.

Practitioners are tutored to avoid being disrupted by passing thoughts and to nudge themselves into concentrating on the breathing once again.

Practitioners may follow four stages:

1. Count the breath at the end of exhalation
2. Count the breath at the beginning of inhalation
3. Focus on the breath without counting
4. Focus only on the spot where breath enters and leaves the nostril/upper-lip area

Stages of Anapanasati: Formally, there are sixteen stages— or lessons — of anapanasati which are divided into four tetrads. The first four steps involve focusing the mind on breathing, which is the body-conditioner (kaya-samkhara in Pali). The second tetrad involves focusing on the feelings (vedan), which are the mind-conditioner (citta-samkhara). The third tetrad involves focusing on the mind itself (citta), and the fourth tetrad focuses on "the truth" (dhamma). (Compare right mindfulness.)

Any meditative session using anapanasati should start at the first stage, continue with the second state, and so on, in sequence. If the session ends at a given stage, then at the next meditative session one should start again at the first stage.

Arupajhana: In Buddhism, the arupajhanas are four successive levels of meditation on non-material objects. These levels are higher than the arupajhanas, and harder to attain. While rupajhanas differ considering their characteristics, arupajhanas differ as their object is determined by the level of the jhana:

- fifth jhana: infinite space,
- sixth jhana: infinite consciousness,

- seventh jhana: infinite nothingness,
- eighth jhana: neither perception nor non-perception.

This has to be understood. In the fourth rupajhana, there is already Upekkha, equanimity and Ekkagata, concentration, but the mind is still focused on a "material" object, as any colour.

- In the fifth jhana, the meditator discovers that there is no object, but only an infinite space, which is empty. This perception motivates the interest of claiming arupajhanas.
- In the sixth jhana, it becomes obvious that space has no existence. There is only infinite consciousness.
- In the seventh jhana appears the feeling that there is no consciousness, but nothingness.
- The eighth jhana consists in the most discrete possible state of mind, which justifies the using of "neither perception nor non-perception".

These "explanations" obviously do not refer to any intellectual, philosophical comprehension, which disappear since the second jhana. They attempt to figure mental process. The arupajhanas are part of the kammatthanas, and are referred to as the four "formless states".

The Two Elements of Arupajhana

Some Tipitaka texts identify arupajhanas as a part of the fourth arupajhana, as they include two elements: upekkha (Sanskrit: upekkha) and ekaggat (Skt: ekgrata).

Upekkha: Upekkha is a Pali word meaning equanimity. The opposition between comfortable sensations and uncomfortable ones disappears. More importantly, it is one of the fourth Jhana's factors, present only in this Jhana.

Ekaggat: Ekaggat or "singlepointedness", as a Jhana's factor, simply means a very deep concentration, which includes the ceasing of stimuli from the exterior world. It is the only jhanic factor present in each Jhana.

Dhyana: Dhyana in Sanskrit or Jhana in Pali refers to a type or aspect of meditation. It is a key concept in Hinduism and

and Buddhism. Equivalent terms are "Zen" in Japanese and "Chan" in modern Chinese.

Dhyana in Buddhism: In the Pali Canon the Buddha describes four progressive states of absorption meditation or Jhana. The Jhanas are said by the Buddha to be conducive to detachment but they must not be mistaken for the final goal of nibbana. The Jhanas are states of meditation where the mind is free from the five hindrances (craving, aversion, sloth, agitation, doubt) and incapable of discursive thinking. The deeper Jhanas can last for many hours. When a meditator emerges from Jhana, his/her mind is empowered and able to penetrate into the deepest truths of existence.

There are four deeper states of meditative absoption called the immaterial attainments. Sometimes these are also referred to as the "formless" Jhanas, or Arupajhana (distinguished from the first four Jhanas, Rupajhana). In the Buddhist canonical texts, the word Jhana is never explicitly used to denote them, but they are always mentioned in sequence after the first four Jhanas.

In East Asia, several schools of Buddhism were founded that focused on dhyna, under the names Chan, Zen, and Seon. According to tradition, Bodhidharma brought Dhyna to the Shaolin temple in China, where it came to be transliterated as "chan" ("seon" in Korea, and then "zen" in Japan).

Jhanas are normally described by the way of the mental factors which are present in these states:

1. Movement of the mind onto the object, Vitakka (Sanskrit: Vitarka)
2. Retention of the mind on the object, Vicra
3. Joy, Piti (Sanskrit: Priti)
4. Happiness, Sukha
5. One-pointedness, Ekaggat (Sanskrit: Ekgrat)
6. Equanimity, Upekkh (Sanskrit: Upekc)

First Jhana (Vitakka, Vicra)

The five hinderances have completely disappeared and intense unified bliss remains. Only the subtlest of mental movement remains-perceiveable only by those who have entered the second Jhana.

Second Jhana (Sukha, Ekaggat)

All mental movement utterly ceases. There is only bliss.

Third Jhana (Sukha, Ekaggat)

One half of bliss disappears (joy).

Fourth Jhana (Upekkh, Ekaggat)

The other half of bliss (happiness) disappears, leading to a state with neither pleasure nor pain, which the Buddha said is actually a subtle form of happiness (more sublime than piti and sukha). The Buddha described the Jhanas as "the footsteps of the tathgata".

Traditionally, this fourth Jhana is seen as the beginning of attaining psychic powers.

Kasina

In Buddhism, kasina are a class of basic visual objects of meditation. There are ten kasina:

1. earth (pathavi kasina),
2. water (apo kasina),
3. air, wind (vayo kasina),
4. fire (tejo kasina),
5. blue, green (nila kasina),
6. yellow (pita kasina),
7. red (lohita kasina),
8. white (odata kasina),
9. enclosed space, hole, aperture (akasa kasina),
10. bright light (aloka kasina).

The kasina are typically described as a colored disk, with the particular colour, properties, dimensions and medium often specified according to the type of kasina. The earth kasina, for instance, is a disk in a red-brown colour formed by spreading earth or clay (or another medium producing similar colour and texture) on a screen of canvas or another backing material.

Kasina meditation is a concentration meditation (variously known in different traditions as samatha, dhyana, or jnana meditations), intended to settle the mind of the practitioner and create a foundation for further practices of meditation. In the early stages of kasina meditation, a physical object is used as the object of meditation, being focused upon by the practitioner until an eidetic image of the object forms in the practitioners mind. In more advanced levels of kasina meditation, only a mental image of the kasina is used as an object of meditation. Unlike the breath, Buddhist tradition indicates that the kasina are not appropriate objects for certain higher levels of meditation, nor for meditation of the vipassana (insight) type.

The ten kasina are part of the forty kammatthana: objects of meditation. They are described in detail by Buddhaghosa in the meditation section of the Visuddhimagga. Unlike breathing meditation, kasina meditation is rarely taught or practiced in the West.

Lotus Position

The Lotus position (Sanskrit: Padmasanam—"lotus posture") is a painful cross-legged sitting posture which originated in representations and meditative practices of Hinduism. It is an established part of the Hindu Yoga tradition, Zen and other meditative disciplines. The position is said to resemble a lotus, encourage "proper" breathing and foster physical stability.

It is perhaps the marquee image of the Eastern religious figure. Famous depictions of the lotus-posture include Shiva, the meditating ascetic god of Hinduism, and Buddha, the founder of Buddhism. The first pictorial representation of padmasana is seen in the ancient Indian depiction of Shiva as Pashupati, Lord of Beasts, in Harrapa.

In this position the right foot is placed on the left thigh and the left foot on the right thigh with the soles of the feet turned up, the spine is straight (note the difference in the position of the feet between this and the "Indian" or "Tailor" style sitting position). In Hinduism the hands are placed on the knees whereas in zazen the hands form the cosmic mudra hand position.

Instructions: "Padmasana (lotus pose)

Sit with the legs straight in front of the body. Slowly and carefully bend one leg and place the foot on top of the opposite thigh. The sole should face upward and the heel should be close to the pubic bone. When this feels comfortable, bend the other leg and place the foot on to top of the opposite thigh. Both knees should ideally touch the ground in the final position.

The head and spine should be held upright and the shoulders relaxed. Place the hands on the knees in chin or jnana mudra. Relax the arms with the elbows slightly bent and check to make sure the shoulders are not raised or hunched. Close the eyes and relax the whole body. Observe the total posture of the body. Make the necessary adjustements by moving forward or backward until balance and alignment are experienced. Perfect alignement indicates the correct posture of padmasana.

Metta

Metta (Pali) or Maitri (Sanskrit) means unconditional and unattached loving kindness. It is one of the ten pramits of the Theravda school of Buddhism, and the first of the four Brahmavihras. The mettabhavana (cultivation of Metta) is a popular form of meditation in Buddhism, practiced with mindfulness of breath, which provides concentration, so as to prevent the loss of compassion.

The object of Metta meditation is to cultivate loving kindness (love without attachment, non-exclusive love) towards all sentient beings. The practice usually begins with the meditator cultivating loving kindness towards themselves (though this is not specifically recommended by the Buddha himself in the relevant suttas/sutras), then their loved ones, friends, teachers, strangers and finally their enemies.

It is a good way to calm down a distraught mind because it is an antidote to anger. Someone who has cultivated Metta will not be easily angered and can quickly subdue anger that arises. They will be more caring, more loving, and more likely to love unconditionally.

Buddhists believe that those who cultivate Metta will be at ease because they see no need to harbour ill will or hostility. Buddhist teachers may even recommend meditation on Metta as an antidote to insomnia and nightmares. It is generally felt that those around a mettful person will feel more comfortable and happy too. Radiating Metta contributes to a world of love, peace and happiness.

Metta Meditation: The Practice of Loving-kindness

Metta signifies friendship and non-violence as well as "a strong wish for the happiness of others", but also less obvious or direct qualities such as showing patience, receptivity, and appreciation. Though it refers to many seemingly disparate ideas, Metta is in fact a very specific form of love – a caring for another independent of all self-interest – and thus is likened to one's love for one's child or parent. Understandably, this energy is often difficult to describe in words; however, in the practice of Metta meditation, one recites specific words and phrases in order to evoke this "boundless warm-hearted feeling." The strength of this feeling is not limited to or by family, religion, or social class. Indeed, Metta is a tool that permits one's generosity and kindness to be applied to all beings and, as a consequence, one finds true happiness in another person's happiness, no matter who the individual is.

Six Stages

The full instructions for the theory and practice of mettabhavana is available in the Visuddhimagga ("The path to purity"), Chapter IX, of the Buddhist scriptures.

The six stages of mettabhavana meditation which are most commonly found involves cultivating loving-kindness towards:

1. Yourself
2. A good friend
3. A 'neutral' person
4. A difficult person
5. All four
6. and then gradually the entire universe

For 2 avoid choosing someone that you feel sexually attracted to, or that is much younger or much older than yourself, or who is dead. For #3 choose someone that you might come in contact with every day, but who does not give rise to strong positive nor strong negative emotions. For #4 traditionally choose "an enemy", but avoid choosing a person who has just wrecked your life, unless you are very well grounded in awareness. For #5 treat them as equals, equally deserving of loving-kindness.

The Metta Meditation is attributed as words of the Buddha (Karaniya Metta Sutta):

This is what should be done
By one who is skilled in goodness,
And who knows the path of peace:
Let them be able and upright,
Straightforward and gentle in speech,
Humble and not conceited,
Contented and easily satisfied,
Unburdened with duties and frugal in their ways.
Peaceful and calm and wise and skillful,
Not proud or demanding in nature.
Let them not do the slightest thing
That the wise would later reprove.
Wishing: In gladness and in safety,
May all beings be at ease.
Whatever living beings there may be;
Whether they are weak or strong, omitting none,
The great or the mighty, medium, short or small,
The seen and the unseen,
Those living near and far away,
Those born and to-be-born—
May all beings be at ease!
Let none deceive another,
Or despise any being in any state.
Let none through anger or ill-will

Wish harm upon another.
Even as a mother protects with her life
Her child, her only child,
So with a boundless heart
Should one cherish all living beings;
Radiating kindness over the entire world.
Spreading upwards to the skies,
And downwards to the depths;
Outwards and unbounded,
Freed from hatred and ill-will.
Whether standing or walking, seated or lying down
Free from drowsiness,
One should sustain this recollection.
This is said to be the sublime abiding.
By not holding to fixed views,
The pure-hearted one, having clarity of vision,
Being freed from all sense desires,
Is not born again into this world

Mindfulness: Mindfulness is the practice whereby a person is intentionally aware of his or her thoughts and actions in the present moment, non-judgmentally. Mindfulness is applied to both bodily actions and the mind's own thoughts and feelings. In Buddhism, the second kind of mindfulness is considered a prerequisite for developing insight and wisdom. Right Mindfulness is the seventh path from the Noble Eightfold Path, which is in its turn the fourth of the Four Noble Truths.

Examples from Meditation and Daily Life: Right mindfulness is also known as Right Meditation. There are many, many forms of mindfulness and meditation. One example of mindfulness is to mentally give a verbal label to each inbreath and outbreath during sitting meditation. So, each time one breathes in, one thinks (*e.g.*) "rising", and each time one breathes out, one thinks "falling". In this type of meditation, the breath serves as a tether that the practitioner uses to bring his or her awareness back to the present moment. By residing more frequently in the present moment, practitioners begin to see both the inner and outer aspects of reality. As one more closely observes inner reality, one finds that happiness is not a quality brought about by a change in outer circumstances, but rather by realizing unhappiness starts with one's own "mindless", and therefore automatic, reactions toward pleasant and unpleasant situations or feelings.

Mindfulness is an activity that can be done at any time. It does not require sitting, or even focusing on the breath, but rather is done by bringing the mind to focus on what is happening in the present moment, while simply noticing the mind's usual "commentary". One can be mindful of the sensations in one's feet while walking, of the sound of the wind in the trees, or the feeling of soapy water while doing dishes. One can also be mindful of the mind's commentary: "I wish I didn't have to walk any further, I like the sound of the leaves rustling, I wish washing dishes wasn't so boring and the soap wasn't drying out my skin", etc. Any activity done mindfully is a form of meditation.

Research and Writing

Largely associated with Buddhism, the practice of mindfulness is also advocated by such people as medical researcher and author Dr. Jon Kabat-Zinn who developed the Mindfulness-Based Stress Reduction (MBSR) programme. MBSR is a form of complementary medicine offered in over 200 U.S. hospitals and is currently the focus of a number of research studies funded by The National Center for Complementary and Alternative Medicine.

Mindfulness is a core exercise used in Dialectical behaviour therapy, a psychosocial treatment Marsha M. Linehan developed for treating people with Borderline Personality Disorder, and is also used in some other newer psychotherapeutical methods, such as Acceptance and Commitment Therapy and Mindfulness-based Cognitive Therapy, which is based on the Mindfulness-Based Stress Reduction programme.

Psychologist Ellen J. Langer has written several books on what she calls "Mindfulness," (including a book with that title), which popularize the experimental work she has done on the topic. Langer acknowledges that what she calls "Mindfulness" is similar to, but distinct from, "Mindfulness" as understood by Buddhism and related traditions.

In addition to various forms of meditation, there are mindfulness training exercises that develop awareness using designated environmental cues. Those who find it difficult to practice sitting meditation may consider trying an approach based on Multiple Intelligences, such as Real-World Mindfulness Training

Rupajhana: In Buddhism, arupajhanas (Sanskrit: arupadhyana "form meditation") are successive levels of meditation in which the mind is focused on a material object: it is a word used in Pali scriptures. Each higher level is harder to reach than the previous one. It is distinguised from arupajhana (Skt: arupadhyana "formless meditation") which is meditation focused on immaterial objects.

There are eight jhanas in total, out of which the first four are arupajhanas. All four arupajhanas are characterized by ekaggata (Skt: ekgrata) which means one-pointedness, *i.e.* the mind focuses singularly on the material object during meditation.

The four arupajhanas are:

1. pamhama-jhana (Skt: prathamadhyana)
2. dutiya-jhana (Skt: dvitiyadhyana),
3. tatiya-jhana (Skt: tetiyadhyana)
4. catuttha-jhana (Skt: caturthadhyana)

These first four jhanas can be characterized by certain factors called jhananga (Skt: dhynanga) whose presence or absence in each arupajhana is summarized in the following table:

jhana	***vitakka & vicra***	***piti***	***sukha***	***ekaggata***	***upekkha***
pamhama-jhana	*	*	*	*	
dutiya-jhana		*	*	*	
tatiya-jhana			*	*	
catuttha-jhana				*	*

The jhananga have the following meanings: vitakka means the noticing of the object of meditation, vicra means the experiencing of the object, piti means rapture, sukha means joy, ekaggata means one-pointedness of concentration, upekkha means equanimity.

Samadhi

Samadhi (Sanskrit, lit. "establish, make firm") is a Hindu and Buddhist term that describes a non-dualistic state of consciousness in which the consciousness of the experiencing subject becomes one with the experienced object, and in which the mind becomes still (one-pointed or concentrated) but the person remains conscious. Sahaj samadhi is the effortless and continual state of perfection of a satguru.

Samadhi is also the Hindi word for a cenotaph, a structure commemorating the dead (akin to a tomb, but without remains).

Etymology: Samadhi IPA: sadmdhi is a Sanskrit term for the state of consciousness induced by complete meditation. Its etymology comes from sam (together or integrated), a (towards), and dha (to get, to hold). Thus the result might be seen to be to acquire integration or wholeness, or truth (samapatti).

Samadhi in Hinduism: Samadhi is the main subject of the first part of the Yoga Sutras called Samadhi-pada. According to Vyasa, a major figure in Hinduism and one of the traditional authors of the Mahabharata, "yoga is samadhi." This is generally interpreted to mean that Samadhi is a state of complete control (samadhana) over the functions and distractions of consciousness.

In practice Samadhi is said to be the state of being aware of one's existence without thinking, in a state of undifferentiated "Beingness." Three intensities (depths) of Samadhi are usually understood in Hinduism.

1. Laja Samadhi
2. Savikalpa Samadhi
3. Nirvikalpa Samadhi (or Sahaja Samadhi)

Laja Samadhi is a latent ("laja"), potential level of samadhi. It begins in deep meditation or trance-even with movement such as dancing. This kind of samadhi is a state of joy, deep and general well-being, and peaceful meditation.

Savikalpa Samadhi refers to the initial temporary state of full-valued samadhi. The conscious mind is still active, as is the kalpa, meaning imagination. One should compare this meaning to that of sankalpa, which is 'wish.' Kalpa takes on a different, but related, meaning to sankalpa because one must use imagination or consciousness (kalpa) to envision a wish or desire (sankalpa). Conversely, vikalpa means "against imagination." At this final level of samadhi, the mind has become quiet and given up its desires and attendant. Vikalpa leads to the Truth, releasing one from any binds of mind (which are mostly imaginations). In Savikalpa Samadhi, we get the taste of Bliss and Beingness, but are still attached to our erroneous identification with the body as well as to our numerous worldly attractions.

Nirvikalpa Samadhi is the end result. There are no more kalpas (imaginings, wishes or other products from work of the mind), because the mind is finally under control. Upon entering Nirvikalpa Samadhi, the differences we saw before have faded and we can see everything as one. In this condition nothing but pure Awareness remains and nothing is missing to take away from Wholeness and Perfection.

Entering samadhi in the beginning takes effort and holding on to a state of samadhi takes even more effort. The beginning stages of samadhi (Laja and Savikalpa Samadhi) are only temporary. By "effort" it is not meant that the mind has to work more. Instead, it means work to control the mind and release the self. Note that normal levels of meditation (mostly the lower levels) can be held automatically, as in "being in the state of meditation" rather than overtly "meditating." The ability to obtain positive results from meditation is much more difficult than simply meditating. It is recommended to find a qualified spiritual master (guru or yogi) who can teach a meditator about the workings of the mind. Samadhi is the only stable unchanging reality; all else is ever changing and does not bring everlasting peace or happiness.

Staying in Nirvikalpa Samadhi is effortless but even from this condition one must eventually return to ego-consciousness. Otherwise, this highest level of Samadhi leads to Nirvana, which means total Unity and the logical end of individual identity (and also death of the body). However, it is entirely possible to stay in Nirvikalpa Samadhi and yet be fully functional in this world. This condition is known as Sahaja Nirvikalpa Samadhi or Sahaj Samadhi (sahaja means "spontaneous" in Sanskrti). Only the truly Enlightened (Satguru) can be and remain spontaneously free.

In Nirvikalpa Samadhi, all attachment to the material world and all karma is dissolved. All awareness is withdrawn step by step from the physical, astral and causal bodies until self-realization or oneness with the soul is achieved. During this process, breathing ceases and the heart stops beating. Aware and fully conscious oneness with soul is then achieved in a most loving way and all cells of physical body are flooded with the Ocean of Divine Love and Divine Bliss for any period of duration-hours, days, weeks until the individual shifts his awareness from the soul back to the physical body. Being fully functional in this world, his awareness stays in connection with the Divine. But some "strange" conditions accompany this state-better health (the body is sustained by Divine Grace), better feelings (even for other people who may contact the body which the enlightened soul has reidentified with) and various miraculous happenings may occur in connection with the Enlightened one.

Maha Samadhi (literally great samadhi) is the Hindi term for a realized yogi's conscious departure from the physical body at death. Which is also known as Nirvana (see above).

Maha Samadhi is the final conscious exit from the physical body. Every infinitesimal piece of attachment or karma is completely surrendered unto God and dissolved into the Divine Ocean of Love. The individual transcends to worlds beyond karma and returns to God, merging into transcendental Bliss.

Samadhi in Bhakti: The Vaishnava Bhakti Schools of Yoga define Samadhi as 'complete absorption into the object of one's love (Krishna)'. Rather than thinking of 'nothing', true samadhi is said to be achieved only when one has pure, unmotivated love of God. Thus even while performing daily activities a practitioner can strive for full samadhi.

Samadhi, a Description: In his book "Autobiography of a Yogi", Paramahansa Yogananda, a reputed, modern day spiritual saint of India and teacher of Kriya Yoga, gives this stirring description of Samadhi consciousness:

My body became immovably rooted; breath was drawn out of my lungs as if by some huge magnet. Soul and mind instantly lost their physical bondage, and streamed out like a fluid piercing light from my every pore. The flesh was as though dead, yet in my intense awareness I knew that never before had I been fully alive. My sense of identity was no longer narrowly confined to a body, but embraced the circumambient atoms. People on distant streets seemed to be moving gently over my own remote periphery. The roots of plants and trees appeared through a dim transparency of the soil; I discerned the inward flow of their sap.

The whole vicinity lay bare before me. My ordinary frontal vision was now changed to a vast spherical sight, simultaneously all perceptive. Through the back of my head I saw men strolling far down Rai Ghat Road, and noticed also a white cow who was leisurely approaching. When she reached the space in front of the open ashram gate, I observed her with my two physical eyes. As she passed by, behind the brick wall, I saw her clearly still.

All objects within my panoramic gaze trembled and vibrated like quick motion pictures. My body, Master's, the pillared courtyard, the furniture and floor, the trees and sunshine, occasionally became violently agitated, until all melted into a luminescent sea; even as sugar crystals, thrown into a glass of water, dissolve after being shaken. The unifying light alternated with materialisations of form, the metamorphoses revealing the law of cause and effect in creation.

An oceanic joy broke upon calm endless shores of my soul. The Spirit of God, I realized, is exhaustless Bliss; His body is countless tissues of light. A swelling glory within me began to envelop towns, continents, the earth, solar and stellar systems, tenuous nebulae, and floating universes. The entire cosmos, gently luminous, like a city seen afar at night, glimmered within the infinitude of my being. The sharply etched global outlines faded somewhat at the farthest edges; there I could see a mellow radiance, ever undiminished. It was indescribably subtle; the planetary pictures were formed of a grosser light.

The divine dispersion of rays poured from an Eternal Source, blazing into galaxies, transfigured with ineffable auras. Again and again I saw the creative beams condense into constellations, then resolve into sheets of transparent flame. By rhythmic reversion, sextillion worlds passed into diaphanous luster; fire became firmament.

I cognized the center of the empyrean as a point of intuitive perception in my heart. Irradiating splendor issued from my nucleus to every part of the universal structure. Blissful amrita, the nectar of immortality, pulsed through me with a quicksilver-like fluidity. The creative voice of God I heard resounding as Aum, the vibration of the Cosmic Motor.

Suddenly the breath returned to my lungs. With a disappointment almost unbearable, I realized that my infinite immensity was lost. Once more I was limited to the humiliating cage of a body, not easily accommodative to the Spirit. Like a prodigal child, I had run away from my macrocosmic home and imprisoned myself in a narrow microcosm.

Samadhi as Leaving the Body: Advanced yogis have been said to consciously leave (or disidentify with) their bodies as a vital step in the attainment of this final samadhi, or soul-liberation. It is at this time that the soul knows a complete and unbroken union with the Heavenly Godhead, and, being free from the limitations of the body, merges effortlessly into the transcendent amrita of Divine Bliss. It is said that sometimes the yogi leaves the body and returns. According to Meher Baba, Jesus entered into nirvikalp samadhi at the time of his crucifiction.

Samadhi in Buddhism: Samadhi, or concentration of the mind, is the third part of the Eightfold Path of the Buddha's teaching: panna(wisdom),, sila (conduct), samadhi or samatha (concentration).

It has been taught by the Buddha using 40 different objects of meditation, such as mindfulness of breathing (anapanasati) and loving kindness (metta). Upon development of samadhi, one's mind becomes purified of defilement, calm, tranquil, and luminous. Once the meditator achieves a strong and powerful concentration, his mind is ready to penetrate and see into the ultimate nature of reality, eventually obtaining release from all suffering. In the language of the eight-fold path, samatha is "right concentration".

Important components of Buddhist meditation, frequently discussed (1, 2) by the Buddha, are the successively higher meditative states known as the four jhanas. The Buddhist suttas mention that samadhi practitioners may develop supernormal powers (called siddhis), and list several that the Buddha developed, but warn that these should not be allowed to distract the practitioner from the larger goal of complete freedom from suffering.

Samatha Meditation

Budhism somewhat considers meditation in general to be an act of concentration on a particular object and/or idea. Therefore, meditation from other religious traditions are soetimes referred to as a variation of samatha meditation that differ in the focus of concentration; such as breathing, scriptural passage, mantra, religious picture, a rock, body (as arepresentation of death), and so on. In this sense, samatha is not a strictly Buddhist meditation.

For Buddhists, it is commonly practiced as a prelude to and in conjunction with vipassana meditation (insight meditation) which focus on Buddhist dharma. Traditionally, there are 40 objects of meditation, although the breath as an object of meditation has always enjoyed the widest popularity. Mindfulness of breathing or napana meditation which accompany Buddhist doctrine of rising and falling can be used for both Samatha and Vipassana Meditation.

Samatha lead to states of meditative absorption or concentration called jhana (Sanskrit: dhyana) which is often translated as state of tranquility or bliss. Thus, it furthers the right concentration aspect of the Noble Eightfold Path. Zamatha is also used in Tibetan Buddhism.

Satipatthana: It is often called the Four Foundations of Mindfulness, namely, mindfulness of the body, of thoughts or the mind, of feelings and of mental objects or qualities. These are also known as the "four frames of reference" or Satipatthana-samyutta. Satipatthana can also be thought of as "establishing mindfulness" or sati-upatthana.

Satipatthana is a way of implementing the right mindfulness and, less directly, the right concentration parts of the Noble Eightfold Path. Satipatthana meditation goes hand-in-hand with vipassana, samatha or calming meditation, and anapana meditation. Satipatthana is practiced with the aims of vipassana and most often in the context of Theravada Buddhism although the principles are also practiced in most traditions of Buddhism which emphasize meditation such as the Zen or Sot Zen tradition.

There is also a Satipatthana Sutta or sutra which is also known as the Discourse on the Foundations of Mindfulness. This sutta explains how to systematically cultivate mindful awareness or sati.

Shikantaza: Shikantaza is literally translated as "only focused on doing sitting". More often it is called: "just sitting" or "silent illumination". It is the main meditation technique of the Soto school of Zen Buddhism. The "just" in "just sitting" has two connotations: Firstly, one should do nothing but the sitting. Secondly, one should sit in the right "mindful" way. The Shobogenzo, an ancient document by the zen master Dogen Zenji (1200-1253), is the most important description of shikantaza.

Meditation Method: Shikantaza is often referred to as the method of no method. While other forms of meditation often have quite clear instructions-such as focusing on the breathing process-shikantaza is held to make use of pure consciousness or mindfulness. The basic idea is that one has to transcend, go beyond, the thinking mind. While sitting one should not think about the sitting, but "really" sit, being one with the sitting. The moment one stops thinking about methods or other things, the correct way suddenly appears. From the viewpoint of Zen, thoughts are a fundamental problem. Most of the time we don't really do things, but rather think about them. That's why zen-masters often stress the point that enlightenment is to be found in things like dish-washing, soaking in a bathtub, and going to the toilet.

There's a famous distinction by Dogen: not-thinking versus non-thinking. One could imagine what it would be like to be "without thoughts", yet a completely blank mind is absolutely not the goal of shikantaza. The thinking has to be transcended altogether. One needs to discover individually what lies behind all these thoughts.

Empirical Explanation of Central Buddhist Concepts: Meditative practice stands in close relationship with central Buddhist concepts. Some people actually "see" them work. The three marks of conditioned existence, a central Buddhist doctrine (see Buddhism), will be explained from the viewpoint of the meditator.

Anicca (impermanence): This first aspect points at the continuous flux of life. Everything is always in motion. Nothing is stable and permanent. Shikantaza is practiced from this point of view. The meditator doesn't discriminate between experiences. Everything is accepted exactly as it is, because ultimately nothing lasts. Knowing there's no real foundation one could hold onto, the actual practice also is of this nature. There's no focusing on any specific experience or thought (as is usual in most meditative practices, for example anapana sati). Because of this the meditation is experienced as a mental state of openness and acceptation.

Dukkha (suffering): This is perhaps one of the most difficult concepts to understand from the perspective of practice. Gautama Buddha says the fundamental problem of life are our desires (see Four Noble Truths). These create attachments. This implicates that we like some, but dislike other experiences. The meditator learns to be fully aware of the experience as it is. It isn't possible to add or subtract anything any more. In this way no new attachment is created and older ones slowly fade away. Over a period of time (some teachers say it takes approximately four years of intensive practice) it is held to be possible to transcend all attachments, thereby acquiring the original and natural state of mind (see enlightenment) in which all experiences reveal themselves as emptiness (sunyata).

Anatta (no self): Buddhism teaches the concept of Anatta. Although every human has the idea of a soul, or a deep feeling of "me" or "self", Buddhism claims there's no stable essence in a human. This can be explained as follows. When incorporating shikantaza in one's life, gradually experiences are welcomed more and more. A great awareness develops itself, which makes it possible to look very carefully at every experience. Which reveals itself as emptiness-as explained before. Then it turns out that even the idea, or this deep "feeling" of a soul every human has, is merely an idea. A thought. Even this idea reveals itself as empty. Which means that the complete identification with this idea no longer holds. Then at once all boundaries break; suddenly there's no difference between the thought of "me" and the sound of a singing bird.

Theoretical Context: For an experienced practitioner Shikantaza is straightforward and simple. Yet, for a beginner this strange "method" is almost impossible to understand. It demands literature study (and/or teaching) and actual practice to understand it thoroughly. Its primary elucidation is in Dogen's Shobogenzo, which is an infamously difficult work of immense philosophical, linguistic, and hermeneutic complexity.

He makes extensive use of paradoxical logic, subversive quotation strategies, imagery, and a rich literary education, and it is impossible to imagine how someone without significant general education and specialized training in the Buddhist textual tradition could begin to access most of the work. Contemporary teaching is often relatively easy to understand, as well as a good introduction to the classic works. Nowadays, many zen-masters have a typical, very "down-to-earth" way of speaking about the practice. Also they try to explain the whole idea of meditation and the connection to this specific practice very lucidly.

Singing Bowl: A singing bowl, also known as a Himalayan bowl or cup gong, is a musical instrument used in Buddhist meditation, dating back many centuries. They are traditionally played in Tibet, Nepal, India, China, Japan, and Korea, mostly by Buddhist monks and in monasteries.

The instrument is made of metal, often including precious metals such as silver and gold. Many other metals, including copper, tin, zinc, lead, iron, and, nickel also feature in many bowls' metallic makeup. Generally, singing bowls can be made of many different metals. Singing bowls are played by hitting, striking, or slowly rubbing/stirring a wooden striker/mallet within or on the top outer surface of the bowl. When done correctly, the bowl will produce a continuous harmonic sound, said to be relaxing and good for the chakra. Some claim that many bowls are "tuned" to a specific chakra. Currently, these bowls are used for meditation, yoga, healing, sound massage, sound therapy, and chakra balancing.

Singing bowls can be found in a variety of shapes, sizes, designs, colours, carvings and craftings. They may be found with various religious themes and symbols, like Tibetan mantra chant om mani padme hum, dragon carvings, mandala carvings, asta mangal (eight auspicious good luck symbols), Buddhist religious symbols and many more. The making of traditional bowls is now considered a lost art. Most old bowls contain very little, if any, ornamentation.

Most bowls are handmade, apart from some tools used for shining and polishing, and fall into two categories. The first is the hand-hammered (beaten) singing bowl, most of which come from Nepal and India, and are manufactured using traditional processes. Hand-hammered bowls are very much more popular for meditation, healing and sound therapy. The second type is the colored, carved and crafted bowl. These are popular for gifts, decorations and souvenir items, and are mostly from Nepal, the Himalayan Country.

Six Yogas of Naropa: The Six Yogas of Naropa describe a set of advanced Tibetan Buddhist tantric meditation practices compiled in and around the time of the Indian monk and mystic Naropa (1016-1100 CE), and conveyed to his student Marpa the translator. The Six Yogas were intended in part to help in the attainment of enlightenment in an accelerated manner.

As Naropa is a Kagyu lineage holder, the six meditative practices are strongly associated with the Kagyu lineage of Tibetan Buddhism. The teachings of Tilopa (988-1069 CE) are the earliest known work on the Six Yogas. Naropa learned the techniques from Tilopa. Naropa's student Marpa taught the Tibetan Milarepa, renowned for his yogic skills. Milarepa in turn taught Gampopa. Gampopa's student, the future first Karmapa, Düsum Khyenpa, attained enlightenment while practicing the Six Yogas.

The Karmapa, the first figure in Tibetan Buddhism to reincarnate, has been strongly associated in certain reincarnations with particular yogic attributes. Many Gelukpa practitioners, including the Dalai Lama, are expert in the Six Yogas of Naropa. The Path to Nirvana In the religion Buddhism, Meditations carry a higher weight in One's practice to attain Nirvana. But, Dana, giving Alms and Shila, observing at least the basic five disciplines (Panchasheela, in Pali) are also vital for One's practice.

In the Maha Satipattana Sutta in Digha Nikaya, the Buddha describes Success of Four Pattana Meditations as: "One who is honest to himself and practice this four Pattana Meditations without a delay, he should be willing to achieve Arahat or Anagami level, in seven days to seven years in time which would ultimately direct to Nirvana"

The Buddha always said: "One who is willing to attain Nirvana, has to understand Four Noble Truth. These Noble Truth is the key to attain Nirvana, without proper understanding of Suffering, Cause of Suffering, Relief of Suffering and the way to end Suffering, These are the four Noble Truth."

Dana, Sila, Samadhi, Prajna

Steps to Nirvana: Dana & Sila practices generate thoughts, (Punya Karma, in Pali), that would create well-being within Samsara. One to give away things, which oneself earned through hard effort might know an easy task. In order to complete Dana Punya Karma, one has to give Alms without a thought of any return.

Observing Sila, or Shiksha, requires much more effort than Dana. Sila is also a Punya Karma, however, basic discipline in oneself would create a powerful platform for such person to attain Nirvana.

The Buddha said: "One who knows that oneself is not breaking these five root disciplines, shall not fear to live anywhere"

Four Satipatthana Meditation: The most important Meditation practices laid by the Buddha are Four Satipatthana Meditations. These four different meditations guide one's mind to understand reality behind the connection between the Mind and the Body. It is increases the Wisdom of dividing Nama and Rupa, Namarupa Pariccheda Guna. By word meaning, Satipatthana means being in Mindfulness.

- Kayanupassana (Contemplation on the Body)
- Vedananupassana (Contemplation on Feelings)
- Cittanupassana (Contemplation on Thoughts)
- Dhammanupassana (Contemplation on Dhammas)

Pattana and Anupassana carries the same meaning. All beings associate some kind of a body, Kya in Pali, and humen associate or link up with a body which is made up with Patawi (Feeling of Hardness), po (Feeling of Liquidity), Thejo (Feeling of hot or mild temperature) and Vayo (Feeling of Air).

Kayanupassana meditation starts from understanding the true nature of body. The body is the beginning of all sarrow, anger, unwise, likeness, and ultimately Suffering.

The Buddha always said:

- "Pagncha Upadanaskadha Dhukkha"
- "One's birth, sickness, death and all affairs along life is suffering. Whatsoever feeling, or thought generated by mind during One's interaction with world or mind through these six bases is Suffering"

Vedana

Vedana is a word in Sanskrit and Pali meaning "feeling" or "sensation".

Synonyms: One of the five aggregates (skandhas). Described by the Buddha as having both mental and physical aspects; therefore vedana offers a means to examine the totality of the mental-physical phenomenon. In the Chain of Conditioned Arising (Pratityasamutpda/Paticcasamuppda), the Buddha explained that tag (tagh), the cause of suffering, arises as a reaction to vedana. By learning to observe vedana objectively one can avoid any new reactions, and can experience directly within oneself the reality of impermanence (anicca). This experience is essential for understanding the nature of attachment, leading to liberation of the mind.

Vipassana

Vipassana (Pali) or vipasyan (Sanskrit) means "insight". While it is often referred to as Buddhist meditation, the practice taught by the Buddha was non-sectarian, and has universal application. It does not require conversion to Buddhism. While the meditation practices themselves vary from school to school, the underlying principle is the investigation of phenomena as they manifest in the five aggregates (skandhas) namely, matter or form (rupa), sensation or feelings (vedana), perception (sanna, Pali sanna), mental formations (samskara, Pali samkhra) and consciousness (vijnna, Pali vinnana). This process leads to direct experiential perception, vipassana.

In a broader sense, vipassana has been used as one of two poles for the categorization of types of Buddhist meditation, the other being samatha (Pali) or Samatha (Sanskrit). Samatha is a focusing, pacifying and calming meditation, common to many traditions in the world, notably yoga. It is used as a preparation for vipassana, pacifying the mind and strengthening the concentration in order to allow the work of insight. This dichotomy is also sometimes discussed as "stopping and seeing." In Buddhist practice, it is said that while samatha can calm the mind, only insight can reveal how the mind was disturbed to start with, which leads to prajna (Pali: panna, knowledge) and jnana (Pali: yanna, pure wisdom) and thus can lead to preventing it from being disturbed again.

The term is also used to refer to a religious movement modelled after Theravada Buddhism which employs Vipassana and panna meditation as its primary techniques and places emphasis on the teachings of the Satipatthana Sutta. Vedana (sensation/feeling) is the primary initial subject of investigation.

Etymology: Vipassana is a Pali word from the Sanskrit prefix "vi-" and verbal root dr. It is often translated as "insight" or "clear-seeing," though, the "in-" prefix may be misleading; "vi" in Indo-Aryan languages is cognate to our "dis." The "vi" in vipassana may then mean to see apart, or discern. Alternatively, the "vi" can function as an intensive, and thus vipassana may mean "seeing deeply". In any case, this is used metaphorically for a particularly powerful mental self-perception. A synonym for "Vipassana" is paccakkha (Pali; Sanskrit: pratyakha), "before the eyes," which refers to direct experiential perception. Thus, the type of seeing denoted by "vipassana" is that of direct perception, as opposed to knowledge derived from reasoning or argument.

Practice of Vipassana: Vipassana meditation is a very simple, logical technique which depends on direct experience and observation. It can be related to the three trainings taught by the Buddha as the basis of a spiritual path-adherence to (abstinence from killing, stealing, lying, sexual misconduct and intoxication), which is not an end in itself but a requirement for the second part, concentration of the mind (samadhi). With this concentrated mind, the third training, in the context of this technique (panna, Sanskrit prajna), is detached observation of the reality of the mind and body from moment to moment. This practice both develops a deep, experiential understanding of the impermanence of reality and also brings to the surface and dissolves deep-seated complexes and tensions.

Put another way, Vipassana meditation consists of the experiential observation of mind and matter (nama and rupa) in their aspects of impermanence, unsatisfactoriness and lack of an inherent, independent essence or self.

To see through the mode of impermanence means to examine things as to whether they are permanent. To see through the mode of unsatisfactoriness means to examine things as to whether they are satisfactory or are imbued with suffering. To see through the mode of non-self means to examine the phenomena that are the objects of the meditation to see if they have a permanent, isolated, and enduring entity. In other words, to see through non-self relates to having a sense of non-doership and a sense of non-possessorship while examining things.

In Vipassana meditation, the meditation object is one's own consciousness, although it can be further refined to be one's consciousness while observing, say, the breath, as in napana meditation. In this context, the modes of seeing refers to focusing on those aspects of consciousness which appear to have (or not have) these characteristics. Some steps are described as vipassana jhanas, or simply as knowledges. A major 19th century vipassana instructor in Burma was Ledi Sayadaw.

Vipassana Today: Today, the term "Vipassana" also refers to a series of meditation techniques used by many branches of modern Theravada Buddhism, for example in modern Sri Lanka, Burma, Laos and Thailand, and to a specific branch of Buddhism popularized by the Indian businessman S.N. Goenka and his mentor U Ba Khin as a nonsectarian form of Buddhism, and also by Americans Joseph Goldstein, Sharon Salzberg, and Jack Kornfield (who were inspired by the monks Mahasi Sayadaw and Ajahn Chah) under the rubric "insight meditation." The vipassana centres supported by S.N. Goenka are present around the world in more than 100 countries. Here locals and visitors can have vipassana training, always given free of charge including accommodation and meals.

Vipassana in the Theravada, Mahayana and Vajrayna

In the Theravada: Vipassana as practiced in the Theravada is the understanding of the Four Noble Truths that were taught by the Buddha. It is understanding the transitory nature of phenomena and the selflessness of persons, that the conceptual consciousness, "I" does not exist.

Most of Theravada's teachers refer to knowledges evolving during practice. The meditator gradually improve his perception of the three marks of existence until he reaches the step sensations constantly disappear, which is called bhaegnupassan nna (Sanskrit: bhaegnupa (yanjnna), knowledge of dissolution.

The yogi will then experience fear and ceasing of attachment, and eventually will reach the step of knowledge of equanimity of formations. This step leads to the attainment of nirvana.

In practise one can use various methods to do Vipassana Meditation. For example one method is that there are 40 topics that can be concentrated by the meditator such as anitya (Pali anicca, impermanence), dukha (Pali dukkha, suffering), roga (illness), and so on. The meditator can meditate on one of these until he sees the truth in everything in the universe.

In the Mahayana: Mahayana Vipasayana consists of meditating on the two truths: conventional truth and absolute truth. One realizes that phenomena likewise have a lack of inherent existence, and have the nature of emptiness (sunyat). This is determined by the inferential path of reasoning and direct observation through meditation.

Gradualism or subitism and the realisation is a debate in the Mahayana. Nevertheless, Huineng, sixth patriarch of the Chan, considered the practice cannot be described as gradualistic nor subitist, but implies people with more or less clear minds.

In the Vajrayna: Mahmudra and Dzogchen use Vipasayana extensively, though in a different manner than in the Theravada. In the Vajrayna (tantric) path, the true nature of mind is pointed out by the guru, and the practitioner takes the path of direct experience.

"In the Sutra path (Theravada) one proceeds by examining and analyzing phenomena, using reasoning. One recognizes that all phenomena lack any true existence and that all appearances are merely interdependently related and are without any inherent nature. They are empty yet apparent, apparent yet empty. The path of Mahmudra is different in that one proceeds using the instructions concerning the nature of mind that are given by one's guru. This is called taking direct perception or direct experiences as the path. The fruition of Samatha is purity of mind, a mind undisturbed by false conception or emotional afflictions. The fruition of vipa[yan is knowledge (prajna) and pure wisdom (jnana). Jnna is called the wisdom of nature of phenomena and it comes about through the realization of the true nature of phenomena."

Vipassana in Prisons: In 1993, Kiran Bedi, a reformist Inspector General of India's prison learned of the success of using Vipassana in a jail in Jaipur, Rajasthan. This 10 day course involved officials and inmates alike. In India's largest prison, Tihar Jail, near New Delhi, another attempt was made. This programme was said to have dramatically changed the behaviour of inmates and jailers alike.

It was actually found that inmates who completed the 10 day course were less violent and had a lower recidivism rate than other inmates. This project was documented in the television documentary, Doing Time, Doing Vipassana. So successful was this programme that it was adopted by correctional facilities in the United States and other countries as well.

Vipassana Jhanas: Vipassana jhanas are steps that describe the evolution of Vipassana's practice. They contrast with samatha's jhanas. The usual description opposes the concentration attained by practicing samatha : the jhanas, and concentration used in vipassana : neighborhood concentration (upacara samadhi).

Nevertheless, Sayadaw U Pandita, a disciple of Mahasi Sayadaw, describes four vipassana jhanas.

- The meditator first explores his body, then his mind, discovering the three characteristics. The first jhana consist in seeing these points and in the presence of vitakka and vicara. Phenomena reveal themselves as appearing and ceasing.
- In the second Jhana, the practice seems quite automatic. Vitaka and vicara both disappear.

- In the third Jhana, piti, the joy, disappears too: there is only happiness (sukha) and concentration.
- Then the practice leads to knowledge. The comfort disappears because the disparition seems the only phenomenon; this step is known as bhanga nupassana. The practice will become very difficult, showing every phenomenon as terrible, dangerous, disenchanting. The desire of freedom will take place. Upekkha, equanimity, will emerge, as the state of mind of an arahant.

The only further steps are the knowledge of attaining Nirvana.

Vipassana Knowledges: This unusual description of Vipassana's practice does not replace vipassana knowledge. Bhanga nupassana is one example of traditional step on the way of wisdom. The steps to nirvana were also described by Buddhaghosa as five purities following purity of discipline (sila) and purity of mind (samadhi), in his famous Visuddhimagga.

Vitakka

Vitakka or Vitarka both in hinduist yoga and Buddhist meditation, means the action of taking care of any object : God, one's body, the Self, a colour, any sensation...

It refers to attention in a different way than cognitive neuroscience, implying a very strong leading of attention, as it leads to more mental process : sukha and ekaggat. Vitakka is the first element to appear in meditation's process, still present in the first Jhana, but absent of any further Jhana.

3

Jain Philosophy

Jain philosophy is based upon eternal, universal truths, according to its followers. Over a period of time, these truths may lapse among humanity and then reappear through the teachings of enlightened humans, those who have reached enlightenment or total knowledge (*Keval Gnan*). Traditionally, in our universe and in our time, Lord Rishabha is regarded as the first to realize the truth, ultimately followed by Lord Parshva (877-777 BCE) and Lord Vardhaman Mahavira.

Jainism teaches that every human is responsible for his/her actions and all living beings have an eternal soul, *jiva*. It insists that we live, think and act respectfully and honour the spiritual nature of all life. Jains view God as the unchanging traits of the pure soul of each living being, chiefly described as Infinite Knowledge, Perception, Consciousness, and Happiness (Anant Gyan, Anant Darshan, Anant Charitra, and Anant Sukh). Jainism does not include a belief in an omnipotent supreme being or creator, but rather in an eternal universe governed by natural laws, and the interplay of its attributes (*gunas*) of matter (*dravys*).

Jain scriptures were written over a long period and the most cited is the *Tattvartha Sutra*, or Book of Reality written by the monk-scholar,Umasvati (or Umasvami), more than 18 centuries ago. The primary figures in Jainism are Tirthankars. Jainism has two main divisions: Digambar and Shvetambar and both believe in ahimsa (or *ahinsa*), asceticism, karma, sansar, and jiva.

Compassion for all life, human and non human, is central to Jainism. Human life is valued as a unique, rare opportunity to reach enlightenment and to kill any person, no matter what crime he may have committed, is unimaginably abhorrent. It is the only religion that requires monks and laity, from all its sects and traditions, to be vegetarian. Some Indian regions have been strongly influenced by Jains and often, the majority of the local non-Jain population has also become vegetarian. History suggests that various strains of Hinduism became vegetarian due to strong Jain influences. In many towns, Jains run animal shelters, *e.g.* Delhi has a bird hospital run by a Jain temple.

A Svetambar Jain worshipping at the Ranakpur temple. The cloth over his face is to prevent spittle or bacteria from from landing on holy images, books or statues of Tirtankars

Jainism's stance on nonviolence goes beyond vegetarianism. Jains refuse food obtained with unnecessary cruelty. Many are vegan due to the violence of modern dairy farms.The orthodox Jain diet excludes most root vegetables, as they believe this destroys life unnecessarily. Another reason for refusing root vegetables is to avoid destroying entire plants. If you eat apples, you do not destroy whole trees, but for root vegetables, whole plants are uprooted. Garlic and onions are avoided as these are seen as creating passion, meaning anger, hatred, jealousy. Observant Jains do not eat, drink, or travel after sunset (which is called Chauvihar) and always rise before sunrise.

Anekantavada, a foundation of Jain philosophy, literally meaning "Nonsingular Conclusivity", or equivalently, "Non-one-endedness". Anekantavad consists of tools for overcoming inherent biases in any one perspective on a topic, object, process, state, or on reality in general. Another tool is The Doctrine of Postulation, Sydvda. Anekantavad is defined as a multiplicity of views for it stresses looking at things from another's perspective.

Jains are traditionally very welcoming and friendly toward other faiths. Several non-Jain temples in India are administered by Jains. The Jain Heggade family has run the Hindu institutions of Dharmasthala, including the Sri Manjunath Temple, for eight centuries. Jains willingly donate money to churches and mosques and usually help with interfaith functions. Jain monks, like Acharya Tulsi and Acharya Sushil Kumar, actively promoted harmony among rival faiths to defuse tension.

Jains have been a palpable presence in Indian culture, contributing to Indian philosophy, art, architecture, sciences, and to Mohandas Gandhi's politics, which led to the mainly non-violent movement for Indian independence.

Universal History and Jain Cosmology

According to Jain beliefs, the universe was never born, nor will it ever cease to exist. Time is divided into Utsurpinis (Progressive Time Cycle) and Avsarpinis (Regressive Time Cycle). An Utsurpini and a Avsarpini constitute one Time Cycle (Kalchakra). Every Utsurpini and Avsarpini is divided into six unequal periods known as Aras. During the Utsurpini half cycle, ethics, progress, happiness, strength, age, body, religion, etc., go from the worst conditions to the best. During the Avsapini half-cycle, these notions deteriorate from the best to the worst. Jains believe we are currently in the fifth Ara of the Avsarpini phase, with approximately 19 000 years until the next Ara. After this Avsarpini phase, the Utsurpini phase will begin, continuing the infinite repetition of the Kalchakra.

When this cycle reaches its lowest level (in the current half-cycle: the sixth Ara), Jainism, as well as all religions will be lost in their entirety. During this time, all humans will be sinners; with short life spans, ugly appearances, and no sense of ethics. Due to this, all individuals during this time will be born in Hell. The world will be a rotting wasteland with little food or water. Then, on the next upswing (start of the Utsurpini half-cycle), the Jain religion will be revived and reintroduced by new [Tirthankar]]s (literally "Crossing Makers" or "Ford Finders"), only to be lost again at the end of the next downswing.

In our era, the twenty-third Tirthankar was an ascetic teacher Parshva, whose traditional dates are 877-777 BC, *i.e.*, 250 years before the liberation of the last Tirthankar, Lord

Mahavir, in 527 BC. Jains regard all Tirthankars as reformers who called for a return to beliefs and practices in accord with the eternal universal philosophy upon which the faith is based. The title Bhagavan ("Lord"), applied to Mahavir and all other Tirthankars, means Venerable.

Bhaktamara Stotra: A Tirthankara is a shelter from ocean of rebirths

The twenty-fourth and final Tirthankar of this Avsarpini is called, Mahvr, the Great Hero (599-527 BC). A wandering ascetic teacher, he recalled Jains to the rigorous practice of their ancient faith.

Jains believe that reality consists of two eternal principles, jiva and ajiva. Jiva consists of infinite identical spiritual units (life); while ajiva (non-jiva) is matter in any form or condition: time, space, and movement.

Both jiva and ajiva are eternal; they are never born or created for the first time and will never cease to exist. The whole world is made up of jivas trapped in ajiva; there are jivas in rocks, plants, insects, animals, human beings, spirits, etc.

Any contact between jiva and ajiva causes the former to suffer and Jains understand that worldly existence inevitably means some suffering. Neither social nor individual reform can totally stop suffering. Every human has jiva which suffers because of its contact with ajiva. To avoid suffering, jiva must leave the four gatis (stages) of Human Life, Heavenly Bodies, Plants/Animals/Insects/Fish Life, and Hell, while remembering the ultimate aim of liberation.

Karma and transmigration keep jiva locked in ajiva. Liberation from the human condition is difficult. Jiva suffers during its infinite reincarnations. Jains believe that every action, good or evil, opens up sense channels (sight, hearing, touch, taste and smell), through which [karma]] adheres to the jiva within, weighing it down and determining the conditions of its next reincarnation.

The consequence of evil actions (Pap) is negative karma, which weighs the jiva down, forcing it to continue reincarnating at low levels. Good deeds (Punya) lead to positive karma, allowing jiva to rise higher in its next life, with less suffering. However, good deeds alone can never lead to liberation.

The fylfot (a.k.a. swastika) is among the holiest of Jain symbols. Worshippers use rice grains to create fylfot around the temple altar.

The way to moksha (release or liberation) is withdrawal from the world. Karma means cause-and-effect and hence every action has inescapable consequences. All Karmas, good or bad, must be shed to attain moksha, since all Karma, good or bad, has consequences and keeps jiva chained in endless reincarnations which lead to suffering to a greater or lesser extent. Liberation warrants prevention and eradication of new karma.

At the end of his life, a liberated soul (in the Jain belief, a Siddha), with no karma to weigh it, will rise free of ajiva, free of the human condition, free of all future embodiments. It will rise to the highest state in the universe, Siddhashila, where jiva, identical with all other pure jivas, will experience its own true nature in eternal stillness, aloneness, liberation and eternal happiness. The way to discard karma is to withdraw from worldly involvement and close the senses and the mind to prevent karma. Such eternal liberation by freeing Jiva from Pudgala (ajiva), so no new reincarnation occurs, is Moksha. Ignorance (ajnana) causes attachment, while true knowledge (keval jnan) leads to liberation. Jains believe that moksha is possible for humans in the fourth Ara of each Utsurpini and Avsarpini.

S. Vernon McCasland, Grace E. Cairns and David C. Yu describe Jain cosmology thus:

> *"In Jain tradition, the first teacher, Rishabh, lived in the third period of Avasarpini, during which half of the world cycle things are getting worse. Since evil had appeared, a teacher / Tirthankara was needed to help people cope with life. In the fourth period, evil proliferated so much that twenty-three more Tirthankaras came into the world to teach people how to defeat evil and achieve moksha. The present time, part of the fifth period, is*

'wholly evil.' Now, men live less than 125 years and the sixth epoch will be worse. 'A man's life span will be only sixteen to twenty years and his height will be reduced to that of a dwarf.... But then the slow upward movement of the first half of the upward cycle, Utsarpini, will begin. There will be steady improvement until, in the first era, man's needs will be fulfilled by wish fullfilling trees, his height will be six miles, and evil will be unknown.' However, eventually things will degenerate, with Avasarpini followed by Usarpini in a neverending cycle." (McCasland, Cairns, and Yu, Religions of the World, *New York: Random House, 1969: pages 485-486)*

The Digambara sect do not believe that women can achieve moksha. The Svetambara sect, actively believes that women can be liberated and a person's gender, or place in the universe is irrelevant to attaining liberation.

Karmic Theory: The Jain religion places great emphasis on the theory of Karma. Essentially, it means that all jivas reap what they sow. A happy or miserable existence is influenced by actions in previous births. These results may not occur in the same life, and what we sow is not limited to physical actions. Physical, verbal, and mental activities play a role in future situations. Karma has long been an essential component of Jainism, and other Indian religions such as Buddhism and Hinduism.

Jainism does not accept the belief that a divine being is responsible for creating pleasant or unpleasant situations. While others may be able to influence one's actions, thus producing Karma, they, as well as one, are bonded to their own Karmas.

Karma, however, is much more complicated than simply 'what goes around comes around'. Though this is not false, it is an simplification: there are many in-depth, comprehensive notions that are also important in the Jain philosophy.

A concept in Jain cosmology and science is the classification and manifestation of Karma. Karmas are constituted of infinitely small "Karman particles". These particles are made up of non-living particles (Pudgals) and are scattered and free-floating across the universe. When the soul acts with a passion such as attachment, anger, deceit, greed, etc., it attracts these. These particles on the soul are [Karma]]. Now, it will only be detached when the Karmic result is reaped, or shed through asceticism.

Depending on the nature of their results, Karmas are divided into two main categories, into eight further categories. The two main categories are Ghati and Aghati karmas. Ghati means destructive or obscuring. The karmas that obscure the true knowledge of the Soul are called Ghati karmas. Karmas that do not obscure the true knowledge of the soul are known as Aghati karmas.

Ghati karmas are divided into four different types of Karma:

1. Knowledge obscuring (Jnanavaraniya) karma: Obscures the full potential of a soul's knowledge. Those with less karmas are more intelligent and learn more easily.
2. Perception obscuring (Darshanavarinya) karma: Diminishes the powers of correct perception through the five senses.
3. Obstructing (Antaraya) karma: Caused by obstructing others' charity, gain or enjoyment.
4. Deluding (Mohaniya) karma: Creates doubt about religion, spiritual teachers and the Jina. It causes attachment and passions such as anger, ego, love, deceit and greed. Of all Karmas, deluding karma is the most difficult to overcome. Once this is eradicated, liberation is assured.

Aghati karmas are of four types:

1. Feeling producing (Vedaniya) karma: Responsible for pain and pleasure.
2. Body Determing (Nam) karma: Determines the physique that the soul will occupy (*i.e.* tall, short, beautiful, ugly, etc.).
3. Status determining (Gotra) karma: Determines social and economic status.
4. Life Span determining (Ayusha) karma: This Karma occurs at the two-thirds point of a jiv's life and

determines the next life. If the jiv is engaged in good deeds, then its next life will be happy. The Jain religion insists that since we do not know that particular point, we should constantly be engaged in religious activities and good deeds. If it is not determined at this point, then the karma will take effect at the two-thirds point of the *remaining* time. If it is still not determined, it will assuredly take place at the moments leading up to, and at death. This karma does not take place if the soul is to be liberated in the current life.

The duration and extent of the results of Karma depends on the soul's vigour and passion when they were acquired. Weak vigour and passion result in smaller, shorter results, while stronger vigour and passion result in greater, longer results.

Nine Tattvas: The backbone of the Jain philosophy, the Nine Tattvas show how to attain salvation. Without knowing them, one cannot progress towards liberation. Jainism explains that Karma theory is intertwined with these nine principles.

Beliefs and Practices: The hand with a wheel on the palm symbolizes the Jain Vow of Ahimsa, meaning non-violence. The word in the middle is "ahimsa." The wheel represents the dharma-chakra, to halt the cycle of reincarnation through relentless pursuit of truth. Jain monks practise strict asceticism and strive to make this, or one of the coming births, their last. The laity, who pursue less rigorous practices, strive to attain rational faith and to do as much good as possible now. Following strict ethics, the laity choose professions that revere and protect life and totally avoid violentlivelihoods.

Jains believe that Devas (angels or celestial beings) cannot help jiva to obtain liberation. This must be achieved by individuals through their own efforts. In fact, devas themselves cannot achieve liberation until they reincarnate as humans and undertake the difficult act of removing karma. Their efforts to attain the exalted state of Siddha, the permanent liberation of jiva from all involvement in worldly existence, must be their own.

The Jain ethical code is taken very seriously. Five vows are followed by both laity and monks/nuns. These are:

1. Nonviolence (ahimsa, or ahinsa)
2. Truth (satya)
3. Non-stealing (asteya)
4. Chastity (brahma-charya)
5. Non-possession or Non-possessiveness (aparigrah)

For laypersons, 'chastity' means confining sexual experiences to marriage. For monks/nuns, it means complete celibacy. Nonviolence involves being vegetarian and some choose to be vegan. Jains are expected to be non-violent in thought, word and deed, towards humans and every living creature. Human life is deemed the highest and it is vital to never harm or upset another. While performing holy deeds, Svetambara Jains wear cloths over their mouths and noses to avoid spittle falling on texts or revered images.

Along with the Five Vows, Jains avoid harboring ill will towards others and practise forgiveness. They believe that Atma (Soul) can lead one to becoming Parmatma (liberated soul) and this must come from one's inner self. No jiv gives another salvation but can only show the way. Jains refrain from all violence (Ahimsa) and have named 18 activities, that should be eradicated:

1. Violence
2. Untruth
3. Theft
4. Unchaste behaviour
5. Possesiveness
6. Anger
7. Arrogance
8. Greed
9. Deceit
10. Attachment
11. Hatred
12. Arguing

13. Accusation
14. Gossip
15. Criticism
16. Prejudice and Dislikes
17. Malice
18. Wrong belief

Mahatma Gandhi was deeply influenced by this Jain emphasis on peaceful, protective living and made it an integral part of his own philosophy.

No description of their use of the two main doors to human spirituality exists, dreams and visions. Or how they relate to or understand reincarnation.

Jain Symbols: The holiest symbol is a simple swastika or svastik. Another incorporates a wheel on the palm of a hand.

Major Jain symbols include:

- 24 Lanchhanas for Tirthankaras
- The Ashtamangals
- Om
- Triratna and Shrivatsa symbols
- A Tirthankar's mother dreams
- Dharma-chakra and Siddha-chakra

Jain Fasting: Fasting is very common among Jains and a part of Jain festivals. Most Jains fast at special times, during festivals, and on holy days. Paryushan is the most prominent festival, lasting 8 days during the monsoon. The monsoon is a time of fasting. However, a Jain may fast at any time, sometimes if s/he feels some error has been committed. Variations in fasts encourage Jains to do whatever they can, to maintain whatever self control is possible for an individual.

The Aim of Fasting: Monks, nuns and laity fast as penance and to control desires. Fasting purifies the body and the mind, reminding one of Mahavir's emphasis on renunciation and asceticism. Mahavir spent months fasting and in contemplation. It is not sufficient simply to stop eating when fasting, s/he must also stop wanting to eat. Control over one's mind is a major goal. If one continues to desire food, the fast is pointless.

There are several fasts:

- Complete fasting: No food or water for a period.
- Partial fasting: Eating less than you desire and to simply avoid hunger.
- Vruti Sankshep: Limiting the number of items eaten.
- Rasa Parityag: Giving up favourite foods.
- Great fasts: Some monks fast for months at a time, following Mahavir, who fasted for over 6 months.

Different Types of Fast:

- Choviyharo Upavas-To give up food and water for the whole day.
- Upavas-To give up only food for the whole day.
- Digamber Upvas-One may drink water only once a day, before sunset.
- Shwetamber Upvas-One may drink Water after Porsi which is boiled and then cooled naturally, provided this is done before sunset
- Ekasan-To eat one meal a day at one sitting and drink water as desired between sunrise and sunset.
- Beasan-To eat two meals a day, (one meal per sitting) and drink water any times between sunrise and sunset.
- Ayambil: Eating food once in one sitting. The food is spice free and boiled or cooked, without milk, curds, ghee, oil, or green/raw vegetables.
- Chathth-To give up both food and water or only food continuously for two days.
- Aththam-To give up food and water or only food continuously for three days.
- Aththai-To give up food and water or only food continuously for eight days.
- Masaksaman-To give up food and water or only food continuously for a whole month.
- Navkarsi: Food and water is consumed forty-eight (48) minutes after sunrise. The orthodox brush teeth and rinse mouths only after sunrise.

- Porsi: Taking food and water three hours after sunrise.
- Sadh-porsi: Taking food and water four hours and thirty minutes after sunrise.
- Purimuddh: Taking food and water six hours after sunrise.
- Avadhdh: Taking food and water eight hours after sunrise.
- Tivihar: After sunset no food or juice shall be taken, but one may drink water. Many Jains practise this daily.
- Navapad oli: During every year for 9 days starting from the 6/7th day in the bright fortnight until the full moon day in Ashwin and Chaitra months, one does Ayambil. This is repeated for the next four and half years. Ayambils may be restricted to one kind of grain per day.

Other austerities are varshitap, Vardhaman, and visasthanak tap, etc.

- Santhara-To give up food and water entirely. This leads to death and is undertaken by someone who has finished all his/her duties and wishes to leave this world peacefully. It gives control over when one dies so everything may be completed and a person may leave of his/her own free will.

Jain Literature: The oldest Jain literature is in Shauraseni and Ardha-Magadhi Prakrit (Agamas, Agama-tulya, Siddhanta texts, etc). Many classical texts are in Sanskrit (Tatvarth Sutra, Puranas, Kosh, Shravakachar, Mathematics, Nighantus etc). Later Jain literature was written in Apabhramsha (Kahas, rasas, grammars, etc), Hindi (Chhahdhal, Mokshamarg Prakashak, etc), Tamil (Jivakacintamani, Kural, etc), Kannada (Vaddaradhane, etc.).

Jain Worship and Rituals: Every day Jains bow and say their universal prayer, the Namaskara Sutra. All good work and events start with this prayer of salutation and worship. Jains have built temples where Tirthankar images are venerated. Jain rituals may be elaborate for symbolic objects are offered and Tirthankars praised in song. But some Jain sects refuse to enter temples or venerate images, considering them simply guides. Sadhumargi Shvetambar Jains, such as the Terapanthi, regard holy statues or temples as totally unnecessary.

Jain Rituals Include:

- Panch-kalyanak Pratishtha
- Pratikraman
- Guru-vandan, Chaitya vandan etc.

Jain marriage ceremonies and family rites are usually variations of orthodox Hindu rituals.

Digambar and Shvetambar Traditions: It is generally believed that the Jain sangha divided into two major sects, Digambar and Shvetambar, about 200 years after Mahvr's nirvan. Some historians believe there was no clear division until the 5th century. The best available information indicates that the chief Jain monk, Bhadrabahu, foresaw famine and led about 12,000 Digambar followers to southern India. Twelve years later, they returned to find the Shvetambar sect and in 453, the Valabhi council edited and compiled traditional Shvetambar scriptures.

Digambar monks do not wear clothes because they believe clothes are like other possessions, increasing dependency and desire for material things, and desire for anything ultimately leads to sorrow. Shvetambar monks wear white clothes for practical reasons and believe there is nothing in Jain scripture that condemns wearing clothes. Sadhvis (nuns) of both sects wear white. These differing views arise from different interpretations of the same holy books. There are minor differences in the enumeration and validity of each sect's literature.

Digambar believe that women cannot attain moksha, while Shvetambars believe that women may certainly attain liberation. Digambars believe that Mahavir was not married while Shvetambars believe the princely Mahavir was married and had a daughter. Apart from doubts about women attaining moksh, another difference is in the first Jain prayer, the

Namaskar Sutra. Other differences are minor and not based on major points of doctrine.

Excavations at Mathura revealed many Jain statues from the Kushana period. Tirthankars are represented without clothes and monks, with cloth wrapped around the left arm, are identified as 'ardha-phalak' and mentioned in some texts. The Yapaniaya sect, believed to have originated from the Ardha-phalak, follows Digambar nudity, along with several Shvetambar beliefs.

Both groups are subdivided into sects, such as Sthanakvasi, Terapanthi, Deravasi, and Bisapanth. Some are 'murtipujak' (image worshippers) while 'non murtipujak', refuse statues or images. Most simply call themselves Jains and follow general traditions rather than specific sectarian practices.

Geographical Spread and Influence

Jain Temple in Ranakpur: The pervasive influence of Jain culture and philosophy in ancient Bihar possibly gave rise to Buddhism. The Buddhists have always maintained that during the time of Buddha and Mahavir, Jainism was already an ancient and deeply entrenched faith and culture in the region. For a discussion about the connections between Jainism and Buddhism see Jainism and Buddhism.

With 5 to 6 million followers, Jainism is among the smallest of the major world religions, but in India its influence is much more than these numbers would suggest. Jains live throughout India; Maharashtra, Rajasthan and Gujarat have the largest Jain population among Indian states. Karnatak and Madhya Pradesh have relatively large Jain populations. There is a large following in Punjab, especially in Ludhiana and Patiala and there were many Jains in Lahore (Punjab's historic capital) and other cities before the Partition of 1947, after which many had to flee to India.

Jains follow their religious practises in Delhi, Mumbai, Kolkata, Chennai, Ahmedabad, Bangalore, and Hyderabad.

There are 85 Jain communities in different parts of India and around the world. They may speak local languages or follow different rituals but essentially, follow the same principles.

Outside India, the United States, United Kingdom, Canada and East Africa (Kenya, Tanzania & Uganda) have large Jain communities. Smaller Jain communities exist in Nepal, Japan, Singapore, Australia, etc. At various times, Jainism was found all over South Asia including Sri Lanka and what are now Pakistan, Bangladesh, Myanmar and Afghanistan.

Jain philosophy and culture have been a major cultural, philosophical, social and political force since the dawn of civilization in Asia, and its ancient influence has been traced beyond the borders of modern India into the Middle Eastern and Mediterranean regions. Jainism is presently a strong faith in the United States and several Jain temples have been built. American Jainism accommodate all the sects.

Over several thousand years, Jain influence on Hindu philosophy and religion has been considerable, while Hindu influence on Jain temple worship and rituals may be observed in certain Jain sects. For a detailed discussion see Jainism and Hinduism.

Jain Contributions to Indian Culture: While the Jains are only 0.4% of the Indian population, their contributions to culture and society in India are considerable. Jainism has influenced Gujarat most significantly. The earliest known Gujarati text Bharat-Bahubali Ras, was written by a Jain monk. The most important people in Gujarat's Jain history were Acharya Hemachandra Suri and his pupil, the Chalukya ruler Kumarpal.

Jains are both among the wealthiest of Indians and the most philanthropic. They run numerous schools, colleges and hospitals and are the most important patrons of the Somapuras, the traditional temple architects in Gujarat.

Jains have greatly influenced Gujarati cuisine. Gujarat is predominantly vegetarian, and its food has pleasing, soothing aromas as pungent onions and garlic are omitted.

In contrast to some religious groups, Jains encourage their monks to do research and go for higher education. Jain monks and nuns, particularly in Rajasthan, have published numerous research monographs. This is unique among Indian religious groups, and parallels Christian clergy. The 2001 census states Jains are India's most literate community and India's oldest libraries at Patan and Jaisalmer are preserved by Jain institutions.

Literature: Jains have contributed to India's classical and popular literature. In Kannada almost the entire early literature is of Jain origin.

- Some of the oldest known books in Hindi and Gujarati were written by Jain scholars.
- Several Tamil classics are written by Jains or with Jain beliefs and values as the core subject.
- Practically all the known texts in the Apabhramsha language are Jain works.

Jainism and Other Religions: South Asia has a rich history of diverse philosophies. Connections among these are discussed at:

- Jainism and Hinduism
- Jainism and Buddhism
- Jainism and Sikhism

Even though Jainism is of Indian origin, it shared some principles with the Hellenic tradition, specially with Stoic and Pythagorean philosophies of Europe. A comparison with modern western religions can be found at:

- Jainism and Christianity
- Jainism and Judaism
- Jainism and Islam

Jiva: In Hinduism and Jainism, a jiva is the immortal essence of a living being (human, animal, fish or plant etc...), it also sometimes referred to the atma, or 'the true self'. It is similar, but not necessarily identical to, the concept of the *soul* as presented in Abrahamic religions. As with all definitions of the 'self', the exact nature of the *jiva* is often a much debated topic.

Definition: In the Bhagavad Gita of Hinduism the jiva is described as immutable, eternal, and indestructible. It is said not to be a product of the material world, but of a higher 'spiritual' nature. At the point of physical death the jiva takes a new physical body depending on karma and the individual desires of the particular jiva in question.

Goals: Jain and Hindu scriptures describe the ultimate goal of the jiva as being either of the following (depending on the particular philosophical tradition):

- 'liberation' from material existence (moksha)
- obtaining pure love of God (bhakti)
- or becoming liberated from the happiness and distress of the world, while still being existant within it (jivanmukta).

Ajiva: In Jainism ajiva is anything that has to do with the nonspiritual in life, the polar opposite of jiva.

Karma: Karma (Sanskrit: from the root k[, "to do", [meaning deed] meaning action, effect, destiny) is a term that comprises the entire cycle of cause and effect. Karma is a sum of all that an individual has done, is currently doing and will do. The effects of all deeds actively create past, present and future experiences, thus making one responsible for one's own life, and the pain in others. In religions that incorporate reincarnation, karma extends through one's present life and all past and future lives as well.

The "Law of Karma" is central in Hinduism, Sikhism, Buddhism, & Jainism. (These religions were formed in India). All living creatures are responsible for their karma-way of life-and for their release from samsara. As a term, it can be traced back to the early Upanishads.

The Law of Karma is taught in the esoteric Christian tradition, Essenian and later Rosicrucian, as the "Law of Cause and Consequence/Effect". However, this western esoteric tradition adds that the essence of the teachings of Christ is that the law of sin and death may be overcome by Love, which will restore immortality.

The process view of release (moksha) from ego-consciousness (ahamkar) through individual responsibility for the totality of action with its inherent karma can be contrasted with the soteriological view of mainstream denominations of Christianity: grace given by faith in the suffering, death and resurrection of a singular saviour.

The Theory of Karma: Karma literally means action. We first take an intent and follow it up with action in the form of thought, speaking to others, planning and execution. According to the Theory of Karma-every action taken to achieve a result or profit (fruit of action) leaves a trace or a karmic residue behind it. It is this residue that keeps accumulating and takes away one's well-being and freedom. Only a person who acts to produce results without worrying about profit (Nishkama Karma or Niskarma) does not accumulate such a Karmic residue. Such action normally comes from love or pure passion.

Quoting from the Bhagavad Gita, Lord Krishna says:

> *"I shall now describe nishkama karma yoga, the path of selfless action. You have a right to perform your natural prescribed duties, but you are not entitled to any fruits of that action. You should neither act with desire to enjoy the fruits of your work, nor, as a result, should you be attached to neglecting your duties."*

So how does Karmic residue work? Well, according to the theory any action whose motive is other than love or passion begins a process of psychological compensation which is difficult to complete. For example, if we helped somebody when we did not want to, then no amount of compensation helps later as we always feel incomplete. Any incomplete action somehow begins to consume a part of our free attention on how to get rewarded. During our journey of life we accumulate considerable Karmic residue and this depletes our free attention significantly. Lack of free attention means lack of intelligence. Many of the bad situations we get into in life are as a result of a lack of intelligence and attention..

Analogs of Karma: If we accept the basic ethical purpose of Karma is to behave responsibly, and the tenet of Karma may be simply stated 'if you do good things, good things will happen to you-if you do bad things, bad things will happen to you', then it is possible for us to identify analogs with other religions that do not rely on Karma as a metaphysical assertion or doctrine.

Karma does not specifically concern itself with salvation-it is just as important within a basic socio-ethical stance. However, as a mechanic, Karma can be identified *in purpose* with the concept of God's relation to 'good works' as found within organized religion, as well as any other religions that assert an omniscient, omnipotent judge, as Hinduism considers with respect to the role of Karma.

Similarly, the Egyptian goddess Maàt (the divine judge) played a similar and impartial role meting out justice in a manner very similar to Karma; Maat could not be appeased by faith or regret-an action done was done, with no space for the more recent theistic concepts of grace, as Hinduism allows for its role of God.

Western Interpretation: An academic and religious definition was mentioned above. Millions of people believe in it and is a part of many cultures and the psyches of millions of people. Others without religious backgrounds, especially in western cultures or with Christian upbringings, become convinced of the existence of Karma. For some, karma is a more reasonable concept than eternal damnation for the wicked. Spirituality or a belief that virtue is rewarded and sin creates suffering eventually leads to a belief in Karma.

According to Karma, performance of positive action results with the reaction of a good conditioning in one's experience, whereas a negative action results in a reaction of a bad response. This may be an immediate result following the act, or a delayed result occurring either in the present life or the next. Thus, meritorious acts may create rebirth into a higher station, such as a superior human being or a godlike being, while evil acts result in rebirth as a human living in less desirable circumstances, or as a lower animal. Some observers have compared the action of karma to Western notions of sin and judgment by God or gods, while others understand karma as an inherent principle of the universe without the intervention of any supernatural Being. In Hinduism, God does play a role and is seen as a dispenser of karma; see Karma in Hinduism for more details. The latter understanding is accurate with regard to Buddhism and Jainism.

Most teachings say that for common mortals, having an involvement with Karma is an unavoidable part of day-to-day living. However, in light of the Hindu philosophical school of Vedanta, as well as Gautama Buddha's teachings, one is advised to either avoid, control or become mindful of the effects of desires and aversions as a way to moderate or change one's karma (or, more accurately, one's karmic results).

Some people have problems with the teaching on karma, often of what exactly the Buddha is asking them to believe in when asking them to have conviction in karma.

- First, action really is happening — it is not an illusion.
- Second, you really are responsible for your actions. There is no outside force like the stars or some good or evil being acting through you. When you are conscious, you are the one who decides what to do.
- Third, your actions have results — you are not just writing on the water — and those results can be good or bad depending on the quality of the intention behind the act.

In the popular American Series My name is Earl, Earl (played by Jason Lee) becomes the victim of karma but eventually realises that he must make up for all the bad things he's done in order to stop being punished.

Spiritism: In Spiritism, karma (termed "the law of cause and effect") plays a central role in determing how one's life needs to be. Spirits are encouraged to choose how (and when) to suffer retribution for the wrong they did in previous lives and so they do. Disabilities, physical or mentail impairment or even an unlucky life are due to the choices the spirit makes before incarnating (*i.e.* being born to a new life).

What sets Spiritism apart from the traditional view is that it understands karma as a condition inherent to the spirit, either incarnated or not: the consequences of the crimes commited by the spirit last beyond the physical life and cause him (moral) pain in the afterlife. The choice of a life of hardships is, therefore, a way to get rid of the pain caused by moral guilt and to perfect qualities that are necessary for the spirit to progress to a higher form. Because Spiritism always accepted the plurality of inhabited worlds, his concept of karma became consideraly complex. There are worlds that are "primitive" (in the sense that they are home to spirits newly born and still very low on intellect and morals) and a succession of more and more advanced worlds to where spirits move as they are elevated. A spirit may choose to be born on a world inferior to his own as a penance or as a mission.

New Age and Theosophy: The idea of karma was popularized in the Western world through the work of the Theosophical Society. Kardecist and Western New Age reinterpretations of karma frequently cast it as a sort of luck associated with virtue: if one does good or spiritually valuable acts, one deserves and can expect good luck; conversely, if one does harmful things, one can expect bad luck or unfortunate happenings. In this conception, karma is affiliated with the Neopagan *law of return* or *Threefold Law,* the idea that the beneficial or harmful effects one has on the world will return to oneself.

There is also the metaphysical idea that, because karma is a force of nature and not a sentient creature capable of making value judgments, karma isn't about good and evil deeds, since applying those labels would require those judgments, but about positive and negative *energy*, where negative energy can include things not seen as "being bad" like sadness and fear, and positive energy can be caused by being creative and solving problems as well as by exuding love and doing virtuous acts. It is referred to as "omniverse karma" or "omni-karma" because it requires the existence of an omniverse (a space which contains all possible universes), and includes concepts such as souls, psychic energy, synchronicity (a concept originally from renowned psychologist Carl Jung, which says that things that happen at the same time are related), and ideas from quantum and theoretical physics.

Reincarnation: Reincarnation, literally "to be made flesh again", as a doctrine or mystical belief, holds the notion that some essential part of a living being (or in some variations, only human beings) can survive death in some form, with its integrity partly or wholly retained, to be reborn in a new body. This part is often referred to as the Spirit or Soul, the 'Higher or True Self', 'Divine Spark', 'I' or the 'Ego' (not to be confused with the ego as defined by psychology).

In such beliefs, a new personality is developed during each life in the physical world, based upon past integrated experience and new acquired experiences, but some part of the being remains constantly present throughout these successive lives as well. It is usually believed that there is interaction between predeterminism of certain experiences, or lessons intended to happen during the physical life, and the free-will action of the individual as they live that life.

This doctrine is a central tenet within the majority of Indian religious traditions such as Yoga, Vaishnavism, Jainism and Sikhism. Buddhist concept of Rebirth, a major part of Buddhist philosophy, differs from the Vedic based traditions, teaches that our "self" (soul) does not reincarnate (see below). Many modern Pagans also believe in reincarnation as do some new Age movements, along with followers of Spiritism, practitioners of certain African traditions, and students of esoteric philosophies.

Overview: Belief in reincarnation is an ancient phenomenon; in various guises humans have believed in a future life since the ancient Egyptians, perhaps earlier, and ancient graves containing both ,people and possessions may testify to beliefs that a person would have need for their treasured possessions once again despite physical death.

In brief, there are several common concepts of a future life. In each of them either the person, or some essential component that defines that person (the soul or spirit) persists in continuing existence:

- Reincarnation in human form.

Successive lives on earth, usually including a belief in a passage through the spiritual world or inner planes between death and rebirth. This is the most common use of reincarnation (also called "rebirth"). In many versions, eventually there is the potential to escape the cycle, *e.g.* by joining God, or achieving enlightenment, some kind of self-realization, a spiritual rebirth, entering a spiritual realm, etc.

- Transmigration of the soul

Beings die, and are returned to this or another existence continually, their form upon return being of a 'higher' or 'lower' kind depending upon the virtue (moral quality) of their present life. This presupposes interchange between human and animal souls, at a minimum; plants and stone may be included, as well.

- Last Day

People live only one life. After death, they may return to the earth or be revived in some final Judgement, or at some final battle (eg the Norse Ragnarok). They may go to heaven or hell at that time, or live again and repopulate the earth. This is an apocalyptic vision of the future.

- Related to these, but no longer including a return to embodied form, are beliefs in an Afterlife

People live on this earth, and then live in some kind of afterlife for the rest of eternity-variously called heaven (paradise) or hell, or the Kingdom of the Dead, or some higher plane, or similar. They do not return to earth as such.

Beliefs in reincarnation or transmigration are widespread amongst religions and beliefs, some seeing it as part of the religion, others seeing in it an answer to many common moral and existential dilemmas, such as "why are we here" and "why do bad things sometimes appear to happen to good people". Reincarnation is therefore a claim that a person has been or will be on this earth again in a different body. It suggests that there is a connection between apparently disparate human lifetimes, and (in most cases) that there may even be covert evidence of continuity between different people's lifetimes, if looked for. Proponents claim this is indeed the case, whilst critics tend to reject the notion due to its metaphysical implications or non-acceptance by science due to other possible explanations of the phenomenon not yet eliminated from consideration. Such evidence tends to be of three kinds:

- Tradition commonly holds that certain people (such as the Dalai or Panchen Lamas in Buddhism) can be identified by looking for a child born at the time of their death, and by certain signs and knowledge that such a child has of their predecessor life beyond the norm. In the case of Buddhism there are well defined tests of such a child.

- In Western culture, regression or near death experience has at times provided what are claimed to be past life memories, some of which can in theory be verified, and some of which might be tested for fraudulent claims. Some aspects of these tend to be quite consistent in some ways (beings of light, messages of love and peace, etc), a factor which to some people lends credence to the idea, and to others supports that "something" is going on but without certainty what that might be.
- Last, for many people, the evidence is internal and empirical, personal belief or experience. This may not be proof as such, but to them, qualifies as sufficient evidence to believe it.

Though some claims to recall past lives have been documented and tested in a scientific manner, mainstream science does not accept reincarnation as a proven phenomenon.

Reincarnation in Various Religions, Traditions and Philosophies: Philosophical and religious beliefs regarding the existence or non-existence of an enduring 'self' have a direct bearing on how reincarnation is viewed within a given tradition. There are large differences in philosophical beliefs regarding the nature of the soul (also known as the jiva or atma) amongst the Dharmic Religions such as Hinduism and Buddhism. Some schools deny the existence of a 'self', while others claim the existence of an eternal, personal self, and still others say there is neither self or no-self, as both are false. Each of these beliefs has a direct bearing on the possible nature of reincarnation, including such concepts of samsara, moksha, nirvana, and bhakti.

In Jainism, particular reference is given to how devas (gods) also reincarnate after they die. A Jainist, who accumulates enough good karma, may become a deva; but, this is generally seen as undesirable since devas eventually die and one might then come back as a lesser being. This belief is also commonplace in a number of other schools of Hinduism.

Classical Greek Philosophy: Some ancient Greek philosophers believed in reincarnation; see for example Plato's *Phaedo* and *The Republic*. Pythagoras was probably the first Greek philosopher to advance the idea.

We do not know exactly how the doctrine of metempsychosis arose in Greece; most scholars do not believe it was borrowed from Egypt or that it somehow was transmitted from ancient Hindu thinkers of India. It is easiest to assume that earlier ideas which had never been extinguished were utilized for religious and philosophic purposes. The Orphic religion, which held it, first appeared in Thrace upon the semi-barbarous north-eastern frontier. Orpheus, its legendary founder, is said to have taught that soul and body are united by a compact unequally binding on either; the soul is divine, immortal and aspires to freedom, while the body holds it in fetters as a prisoner. Death dissolves this compact, but only to re-imprison the liberated soul after a short time: for the wheel of birth revolves inexorably.

Thus the soul continues its journey, alternating between a separate unrestrained existence and fresh reincarnation, round the wide circle of necessity, as the companion of many bodies of men and animals." To these unfortunate prisoners Orpheus proclaims the message of liberation, that they stand in need of the grace of redeeming gods and of Dionysus in particular, and calls them to turn to God by ascetic piety of life and self-purification: the purer their lives the higher will be their next reincarnation, until the soul has completed the spiral ascent of destiny to live for ever as God from whom it comes. Such was the teaching of Orphism which appeared in Greece about the 6th century BC, organized itself into private and public mysteries at Eleusis and elsewhere, and produced a copious literature.

The earliest Greek thinker with whom metempsychosis is connected is Pherecydes; but Pythagoras, who is said to have been his pupil, is its first famous philosophic exponent. Pythagoras probably neither invented the doctrine nor imported it from Egypt, but made his reputation by bringing Orphic doctrine from North-Eastern Hellas to Magna Graecia and by instituting societies for its diffusion.

The real weight and importance of metempsychosis in Western tradition is due to its adoption by Plato. Had he not embodied it in some of his greatest works it would be merely a matter of curious investigation for the Western anthropologist and student of folk-lore. In the eschatological myth which closes the Republic he tells the story how Er, the son of Armenius, miraculously returned to life on the twelfth day after death and recounted the secrets of the other world. After death, he said, he went with others to the place of Judgment and saw the souls returning from heaven and from purgatory, and proceeded with them to a place where they chose new lives, human and animal. He saw the soul of Orpheus changing into a swan, Thamyras becoming a nightingale, musical birds choosing to be men, the soul of Atalanta choosing the honours of an athlete. Men were seen passing into animals and wild and tame animals changing into each other.

After their choice the souls drank of Lethe and then shot away like stars to their birth. There are myths and theories to the same effect in other dialogues, the Phaedrus, Meno, Phaedo, Timaeus and Laws. In Plato's view the number of souls was fixed; birth therefore is never the creation of a soul, but only a transmigration from one body to another. Plato's acceptance of the doctrine is characteristic of his sympathy with popular beliefs and desire to incorporate them in a purified form into his system. Aristotle, a far less emotional and sympathetic mind, has a doctrine of immortality totally inconsistent with it.

In later Greek literature the doctrine appears from time to time; it is mentioned in a fragment of Menander (the Inspired Woman) and satirized by Lucian (Gallus 18 seq.). In Roman literature it is found as early as Ennius, who in his Calabrian home must have been familiar with the Greek teachings which had descended to his times from the cities of Magna Graecia. In a lost passage of his Annals, a Roman history in verse, Ennius told how he had seen Homer in a dream, who had assured him that the same soul which had animated both the poets had once belonged to a peacock. Persius in one of his satires (vi. 9) laughs at Ennius for this: it is referred to also by Lucretius (i. 124) and by Horace (Epist. II. i. 52). Virgil works the idea into his account of, the Underworld in the sixth book of the Aeneid (vv. 724 sqq.). It persists in antiquity down to the latest classic thinkers, Plotinus and the other Neoplatonists.

Judaism and Kabbalah: Classic works of the Kabbalah, *Shaar ha Gilgulim* ("Gate of Reincarnations") of Arizal or *Isaac Luria*, describes complex laws of reincarnation *gilgul* and impregnation *ibbur* of 5 different parts of the soul. It shows many references of reincarnation in the Hebrew Bible (the Tanach). The notion of reincarnation is not openly mentioned in the Hebrew Bible. The classical rabbinic works (midrash, Mishna and Talmud) also are silent on this topic.

The concept was elucidated in an influential mystical work called the *Bahir* (Illumination) (one of the most ancient books of Jewish mysticism) which was composed by the first century mystic Nehunia ben haKana, and gained widespread recognition around 1150. After the publication of the *Zohar* in the late 13th century, the idea of reincarnation spread to most of the general Jewish community.

While ancient Greek philosophers like Plato and Socrates attempted to prove the existence of reincarnation through philosophical proofs, Jewish mystics who accepted this idea did not. Rather, they offered explanations of why reincarnation would solve otherwise intractable problems of theodicy (how to reconcile the existence of evil with the premise of a good God.)

Rabbis who accepted the idea of reincarnation include the founder of Chassidism, the Baal Shem Tov, Levi ibn Habib (the Ralbah), Nahmanides (the Ramban), Rabbenu Bahya ben Asher, Rabbi Shelomoh Alkabez and Rabbi Hayyim Vital. The argument made was that even the most righteous of Jews sometimes would suffer or be murdered unjustly. Further, children would sometimes suffer or be murdered, yet they were obviously too young for them to have committed sins that God would presumably punish them for. Jewish supporters of reincarnation said that this idea would remove the theodicy: Good people were not suffering; rather, they were reincarnations of people who had sinned in previous lifetimes. Therefore any suffering which was observed could be assumed to be from a just God. Yitzchak Blua writes "Unlike some other areas of philosophy where the philosophic battleground revolves around the truth or falsehood of a given assertion, the *gilgul* debate at points focuses on the psychological needs of the people."

Martin Buber's collection of *Legend of the Baal-Shem* (*Die Chassidischen Bücher*) includes several of the Baal Shem Tov's stories that explicitly discuss concrete cases of reincarnating souls.

Rabbis who rejected the idea of reincarnation include Saadia Gaon, Hasdai Crescas, Yedayah Bedershi (early 14th century), Joseph Albo, Abraham ibn Daud and Leon de Modena. Saddia, in Emunoth ve-Deoth, concludes Section vi with a refutation of the doctrine of metempsychosis. Crescas writes that if reincarnation was real, people should remember details of their previous lives. Bedershi offers three reasons why the entire concept is dangerous:

1. There is no reason for people to try and do good in this life, if they fear that they will nonetheless be punished for some unknown sin committed in a past life.
2. Some people may assume that they did not sin in their past life, and so can coast on their success; thus there is no need to try hard to live a good life. In Bedershi's view, the only psychologically tenable worldview for a healthy life is to deal with the here-and-now.
3. The idea presents a conundrum for those who believe that at the end of days, God will resurrect the souls and physical bodies of the dead. If a person has lived multiple lives, which body will God resurrect?

Joseph Albo writes that in theory the idea of gilgulim is compatible with Jewish theology. However, Albo argues that there is a purpose for a soul to enter the body, creating a being with free will. However, a return of the soul to another body, again and again, has no point. Leon De Moden thinks that the idea of reincarnation make a mockery of God's plans for humans; why does God need to send the soul back over and over? If God requires an individual to achieve some perfection or atone for some sin, then God can just extend that person's life until they have time to do what is necessary. de Modena's second argument against reincarnation is that the entire concept is absent from the entire Bible and corpus of classical rabbinic literature.

The idea of reincarnation, called *gilgul*, became popular in folk belief, and is found in much Yiddish literature among Ashkenazi Jews. Among a few kabbalists, it was posited that some human souls could end up being reincarnated into non-human bodies. These ideas can be found in a number of Kabbalistic works from the 1200s, and also among many mystics in the late 1500s. A distinction was made, however, between actual Transmigration and this form of reincarnation; the non-human subject had its own soul already, the human soul simply 'rode along with' the rock, or tree, or giraffe waiting to be 'elevated,' that is, to be raised to a higher level and to gradually approach the level of human again.

The cow eats the grass, elevating the soul within it, the soul rides with the cow a while until a person eats the cow, and then the soul is elevated to the max. Rabbi Chaim Vidal, when asked how he came to be the foremost desciple and sole transmitter of the teachings of his teacher, the great Issac Luria, credits, not study or mitzvot, but his diligence in blessing his food: "For this way I elevate the souls therein. These souls then become my witnesses in the Heavenly Realm, and empowerme to receive even greater revelations."

"Over time however, the philosophical teaching limiting reincarnation to human bodies emerged as the dominant view. Nonetheless, the idea that one can reborn as an animal was never completely eliminated from Jewish thought, and appears centuries later in the Eastern European folk tradition". [Simcha Paull-Raphael.

While many Jews today do not believe in reincarnation, the belief is common amongst Orthodox Jews, particularly amongst Hasidim; some Hasidic siddurim (prayerbooks) have a prayer asking for forgiveness for one's sins that one may have committed in this *gilgul* or a previous one.

Gnosticism: Many Gnostic groups believed in reincarnation. For them, reincarnation was a negative concept: Gnostics believed that the material body was evil, and that they would be better off if they could eventually avoid having their 'good' souls reincarnated in 'evil' bodies.

The Gnostic Gospel of the Nazirenes

1. *As Yeshua sat by the west of the temple with his disciples, behold there passed some carrying one that was dead, to burial, and a certain one said to Him, "Master, if a man die, shall he live again?"*
2. *He answered and said, "I am the resurrection and the life, I am the good, the beautiful, the true; if a man believe in me he shall not die, but live eternally. As in Adam all (1997 = are bound to cycles of rebirth) die, so in the Messiah shall all be made alive. Blessed are the dead who die in me, and are made perfect in my image and likeness, for they rest from their labours and their works do follow them. They have overcome evil, and are made pillars in the temple of my God, and they go out no more, for they rest in the eternal."*
3. *"For them that persist in evil there is no rest, but they go out and in, and suffer correction for ages, till they are made perfect. But for them that have done good and attained to perfection, there is endless rest and they go into life everlasting. They rest in the eternal."*
4. *"Over them the repeated death and birth have no power, for them the wheel of the eternal revolves no more, for they have attained to the center, where is eternal rest, and the center of all things is God."*

Note: The text above is not from the original Gospel of the Nazirenes, which now exists only in fragments. Rather, it is the product of "channeling" and of recent origin.

The texts contains several parallels to the Gospels, which are, though, traditionally interpreted differently in their context:

"I am the resurrection and the life; he who believes in me, though he die, yet shall he live, and whoever lives and believes in me shall never die."

Him who overcomes I will make a pillar in the temple of my God. Never again will he leave it.

Bible and Reincarnation: Almost all present official Christian denominations reject reincarnation: exceptions include the Liberal Catholic Church and the Rosicrucian Fellowship. Doctrines of reincarnation were known to the early Church (before the 6th century A.D.), and believers in reincarnation claim that these doctrines were embraced or at least tolerated within the Church at that time. Two Church Fathers, Origen and Clement of Alexandria are frequently cited as supporting this. However, this cannot be confirmed from the existent writings of Origen. He was cognizant of the concept of reincarnation (metensomatosis "re-embodiment" in his words) from Greek philosophy, but he repeatedly states that this concept is no part of the Christian teaching or scripture. He writes in his Comment on the Gospel of Matthew: "In this place [when Jesus said Elijah was come and referred to John the Baptist] it does not appear to me that by Elijah the soul is spoken of, lest I fall into the doctrine of transmigration, which is foreign to the Church of God, and not handed down by the apostles, nor anywhere set forth in the scriptures".

Some reincarnation followers state that Origen's writings have only come down to us heavily edited 'to conform to Church doctrine', and some Origen's writings were later declared heretical by the Church (though Origen himself was not). However, Gregory of Nyssa cites Origen:

By some inclination toward evil, certain souls... come into bodies, first of men; then through their association with the irrational passions, after the allotted span of human life, they are changed into beasts, from which they sink to the level of plants. From this condition they rise again through the same stages and are restored to their heavenly place. (B.W. Butterworth, On First Principles, Book I, Chapter VIII (New York: Harper & Row, 1966).

They also state that before the Church expurged what it considered his heretical ideas from editions of his works, other quotes of Origen were also recorded by early Church fathers that make it clear that he did indeed teach reincarnation. A discussion of Origen's relationship to reincarnation, including many more quotes, can be found at Kevin Williams' Near Death Experiences website.

Kurt Eggenstein claims that "Jerome wrote in a letter to Demetrius that among the early Christians, the doctrine of reincarnation had been passed on to the elect, as an occult tradition." He also gives a quote from Gregory of Nyssa, saying "It is a necessity of nature that the soul becomes purified in repeated lives", though the source and the translation are uncited. His book claims many more Christian authorities supported a belief in reincarnation.

In the New Testament, there are several passages that some people use to demonstrate that a belief in reincarnation was prevalent amongst those of Jesus' inner circle. He is asked if he is Elias, for example, in John 1:21; in Matthew 16:13-14 Jesus asks his disciples, *'Who do men say that I, the Son of Man, am?' And they said, 'Some say that you are John the Baptist; some, Elias; and others, Jeremias, or one of the other prophets.'* According to those people, such statements are only comprehensible if Jesus' disciples believed in reincarnation-although one can argue that this merely confirms that they knew this kind of belief, or that, even if the disciples held such beliefs at that moment, that does not confirm that Christ himself endossed it. Finally, in Matthew 11:13-14, Jesus says: *For all the prophets and the law prophesied until John. And if ye will receive it, this is Elias, which was for to come.*

This can be understood in the light of the traditional Jewish prophecy that Elijah (Elias) would return one day, bringing on the Messianic age. John however denied that he was *literally* Elias John 1:21. And further, Elijah was transfigured and taken up into heaven (2Kings 2:11). Since he did not die, he would have no need of reincarnation to return again as prophesied by Malachi. However, the whole descriptive image of the transfiguration does not state that Elijah did not die, just that *"went up by a whirlwind into heaven"*, and what this process of ascending to heaven implies to the physical body is not known in the mainstream theological study. On the other hand, the orthodox interpretation of Elijah's entering heaven in the physical body would contradict, among other passages, the New Testament's statement that *"flesh and blood cannot inherit the kingdom of God"* (1Corinthians 15:50); unless "heaven" is considered not to be a part of the "kingdom of God" and that concept would contradict, among other passages, the preaching of John the Baptist *"Repent ye: for the kingdom of heaven is at hand."* (Matthew 3:1-2) and the statement of Christ: *"In my Father's house are many mansions"*.

Matthew 19:28 states: "Verily I say unto you, that ye which have followed me, in the regeneration (Greek — pale-genesia literally, rebirth) when the Son of Man shall sit in the throne of his glory, ye also shall sit upon twelve thrones, judging the twelve tribes of Israel." A more well-known passage from John 3:3 reads, "...Except a man be born again (Greek — ano-then), he cannot see the Kingdom of God." The quote from John is sometimes translated as "born from above", and is the inspiration for the modern evangelical movement. Some readers interpret these passages to indicate reincarnation; however, Christian churches read them to refer to baptism or conversion, in a manner similar to what we know today as Born-again Christian.

In John 9:1, the discples put the question to Jesus, regarding a man who was blind from birth, "Master, who did sin, this man, or his parents, that he was born blind?" The disciples appear to be citing two of the most plausible theories of the time: reincarnation, and sins of the parents (or, effects of parenting).

This suggests that reincarnation was known to the disciples. Jesus's answer, "Neither hath this man sinned, nor his parents: but that the works of God should be made manifest in him" is open to interpretation, but it is apparent that he did not rebuke the disciples for suggesting the idea of reincarnation itself. In fact, one could interpret that he tacitly affirmed both hypotheses, while pointing to a third explanation in this particular case.

The Gnostic gospels include clear references to reincarnation, and it is clear that this early Christian (heretical) sect believed in this (see above). In the Gospel of Thomas, Nag Hammadi documents, "The kingdom is like a man who had a hidden treasure in his field without knowing it. And after he died, he left it to his son.

The son did not know (about the treasure). He inherited the field and sold it. And the one who bought it went plowing and found the treasure. He began to lend money at interest to whomever he wished." The "field" can be interpreted as our phenomenal world of sense experience; the "treasure" the essential Self; "inheriting" as reincarnating; and "plowing" as spiritual search and spiritual discipline.

A number of Evangelical and (in the USA) Fundamentalist Christian groups denounce any belief in reincarnation as heretical, and explain any phenomena suggestive of it as deceptions of the devil, using the Bible as source for such claims. In fact, although the Bible never mentions the word *reincarnation*, there are several passages through New Testament that Orthodox Christians interpret as openly rejecting reincarnation or the possibility of any return or contact with this world for the souls in Heaven or Hell (see Hb 9:21 and Luke 16:20-31)

There are various contemporary attempts to reconcile Christianity and reincarnation. See:

- Geddes Macgregor, *Reincarnation in Christianity : A New Vision of Rebirth in Christian Thought*
- Rudolf Steiner, *Christianity and Mystical Fact.*

Islam: Mainstream Islam rejects the concept of reincarnation. Believing in reincarnation into this world, in such a way that it could be interpreted as a denial of resurrection may constitutes apostasy in Islam.

A very few sufi groups belive in reincarnation claiming that this concept is mentioned in Quran 2:28:

"How can you deny God, when you were dead and God gave you life? Then God will cause you to die, and then revive you, and then you will be returned to God." (Quran 2:28)

The mainstream Islam rejects this understanding of the verse, claiming that it refers to the worldly human life and the consequent resurrection in the hereafter.

It is claimed by some sufi groups that the mystics and poets in the Islam tradition have celebrated this belief:

"I died as mineral and became a plant,
I died as plant and rose to animal,
I died as animal and I was man.
Why should I fear?
When was I less by dying?
Yet once more I shall die as man,
To soar with angels blest;
But even from angelhood I must pass on..."

(excerpt from *Masnawi*, by Hazrat)

In his *Masnavi*, Rumi speaks about the *"seventy-two forms I have worn"*.

Modern Sufis who embrace the idea of reincarnation includ Bawa Muhaiyadeen (see his *To Die Before Death: The Sufi Way of Life*) and Hazrat Inayat Khan (see *The Sufi Message*, vol. V, part 3). Reincarnation has also been used to reconcile the Quran's apparent identification of Miriam, the mother of Isa as the sister of Aaron and daughter of Amran, all of whom lived well before the first century CE.

Some verse of Quran that seem to discount repeated lives:

- "From the (earth) did We Create you, and into it Shall We return you, And from it shall We Bring you out once again. (The Quran, 20:55)"
- "'And Allah has produced you from the earth, Growing (gradually), And in the End He will return you Into the (earth), And raise you forth (Again at the Resurrection).' (The Quran, 71:17-18)"
- "Nor will they there Taste Death, except the first Death; and He will preserve Them from the Penalty Of the Blazing Fire. (The Quran, 44:56)"

- "'Is it (the case) that We shall not die, except our first death, And that we Shall not be punished?' Verily this is The supreme achievement! For the like of this Let all strive, Who wish to strive. (The Quran, 37:58-61)"

Native American Nations: Reincarnation is an intrinsic part of many Native American and Inuit traditions. Regardless of the actual religious beliefs and practices of today's Native Americans, with varying religious beliefs, the idea has survived for centuries. In the now heavily Christian Polar North (now mainly parts of Greenland and Nunavut), the concept of reincarnation is enshrined in the Inuit language. The survival of the concept of reincarnation applies across the Nations in varying degrees of integrity. The Nations are, of course, now sandwiched between Eastern [Native] and Western traditions.

Moksha

In Jainism, Moksha and Nirvana are the same. When a soul (atman) achieves Nirvana, it is released from the cycle of births and deaths, and achieves its pure self. It then becomes a Siddha (literally one who has accomplished his ultimate objective), and hence a god, worthy of worship.

In Jainism, attaining Moksha requires annihilation of all karmas, good and bad; because if karma is left, it must bear fruit.

Ahimsa

In Jainism, the ahimsa-vrata, or vow of ahimsa, is the first of the five mahavratas, or great vows. All animal life, and most plant life, is considered sentient. Any action endangering such life, including violence, animal sacrifice, drinking liquor, eating honey, potatoes or certain fruits, and eating at night, is forbidden. Some Jains wear a cloth over their mouths to avoid inhaling airborne life forms.

The ethical code of Jainism is taken very seriously. Summarized in the Five Vows, they are followed by both lay people and monastics. These are:

1. non-injury (ahimsa)
2. non-lying (satya)

3. non-stealing (asteya)
4. non-possession (aparigrah)
5. chastity (brahmcharya)

The Jain conception of ahimsa involves three times three—the three actions (*karanas*) of himsa in the three modes (*yogas*)—observances:

Neither by action or by speech or by thought:

1. do injury to oneself (*krita*)
2. cause injury to be done to others (*karita*)
3. approve injury done by others (*anumata*, *mananat*, or *anumodana*)

Satya

Satya is a Sanskrit word that loosely translates into English as "Truth." Satya is pronounced like Libya, with two and a half or three syllables. It is a term of power due to its purity and meaning and has become the emblem of many peaceful social movements, particularly those centered on social justice, environmentalism and vegetarianism.

Philosophical Meaning: The philosophical meaning of the word 'Satya' is "unchangeable", "that which has no distortion", "that which is beyond distinctions of time, space, and person", "that which pervades the universe in all its constancy". Human life progresses through different stages — from childhood to adolescence, from adolescence to youth, and youth to old age. It is through these changes that people progress in the manifest world. That is why human life or its receptacle, the body, is not Satya' or Truth.

Subtle Meaning: There is a more subtle meaning of the word, 'Satya', which is Citsvaru'pa (the Supreme consciousness) or Purus'a. In the field of Sa'dhana' or intuitional practice, the meaning of 'Satya' is 'Parahit'artham' va'unmanaso yatha'rthatvam' satyam *i.e.*, Satya is the benevolent use of words and the mind for the welfare of others. This is to say that a benevolent sage must be truthful regardless of the meaning of satya.

Common Interpretation: The accepted interpretation however, is "the Truth which equals love." This concept of truth is not merely a synonym of fact or correctness, but is more metaphysical, like the difference between "brain" and "mind." This 'bigger picture' notion of truth, at least as far as the term Satya is concerned, implies a higher order, a higher principle, a higher knowledge, but not necessarily a higher being or creator. Satya is what one becomes aware of upon becoming a Bodhi (enlightened or awakened person—'Buddha' means awakened one).

Thus, this topic is an aspect more akin to the sum of the rules of the universe—the 'universal reality.' This idea of a universal reality is common in Eastern philosophy. Combined with other words, Satya acts as modifier, like "ultra" or "highest," or more literally "truest," connoting purity and excellence. Examples: Satyaloka = highest heaven; Satya Yuga = the original, "golden" and best of the four cyclical cosmic ages in Hinduism. Note, in Hinduism, we are currently in the Kali Yuga, the most depraved, degraded, and corrupt of the four cycles, which began on 3102 BC and will end and be replaced by the Satya Yuga in another 427,000 years.

Brahmacharya

Usage: This word in one common usage generally denotes a particular phase of life (first 20-25 years of Life). Ancient Hindu culture divides the life in 100 year life span. This is the stage when the young child leads a student life, (ideally in the Gurukula-the household of the Guru). This stage of life is preceded by the Upanayanam of the child, during which ceremony, they are considered to be born again. It is the first in the four phases of human life, namely, Brahmacharya, Grihastha, Vanaprastha, and finally Sanyasa, as prescribed by Manusmriti for the Dhwija castes, in the Hindu system of life. The practise of brahmacharya among other things requires one to be celibate.

The word, Brahmacharya, is also used for the vow of celibacy that a Hindu *sanyasi* or renunciate may take at any age when they have understood that living for material/sensual pleasures will never bring the perfect happiness that their soul desires. Thus their life becomes centered on surrender to Guru and God with the firm hope of God realization and the perfect Divine Happiness we all desire.

Yoga: In yoga, the term can also mean disciplining and preservation of sexual energy and is discussed in Patanjali's Yoga Sutras. There are many yogic techniques, such as meditation and asanas (*e.g.* shirsasana) that can help one to achieve this aim.

Diet and Ahimsa: Brahmacharya is also observed to contain one's sensual desires for food and taste, as well as materialism. Most brahmacharis prescribe to avoiding the consumption of meat, spices and cooked foods, said to cultivate the taste buds and pleasure senses of the mind.

To terminate egotism and possessiveness, as a part of curbing one's aggressiveness and tendency to commit violence, one must cease to covet all material possessions.

Modern Brahmacharis: Most Hindu, Buddhist and Jain monks take the vow for life, committing themselves to work of religious service and study. Mahatma Gandhi, the great Indian political and spiritual leader had embraced the vow and lifestyle permanently at age 36.

Many brahmacharis have the final goal of nirvana, or moksha in mind when they pursue strictly disciplined lifestyles.

Tripadi

The Three Pronouncements: Shraman Bhagwan Mahavir expounded and established the Dharma (the Jain philosophy) and communicated it to his first disciple, Indrabhuti Gautam Swami and ten other Ganadhars (Chief disciples) in three statements which constitute the foundation of the Jain philosophy; and a comprehensive statement of its essentials. These three statements are known as Tripati.

(1) Upanneyi Va-There is production of matter: [Utpad = Utpatti = production.]

(2) Vigameyi Va-Matter gets destroyed: [Vyav=/aya = destruction.]

(3) Dhuveyi Va-Matter is permanent: [Dhrauvya = Sthiti = permanent existence.]

Sat is Dravya. It is matter. Though it may appear in various forms, it is permanent in its original and essential form. It is permanent. It remains in its original form. It is called Padarth or substance. This matter assumes various transformations. It has its original form as well as its changed forms.

Changes are taking place every moment in matter and in every aspect of it. Production and destruction are endless processes. But on account of these changes, Matter does not experience any loss in its original form.

Matter as Dravya remains permanent and sound. But changes occur and changed forms are destroyed. For this reason, the Jain dharma does not consider any matter either as always permanent or as always transitory. When we notice the destruction of substance. it is not really loss or destruction but only a change or a transformation of matter. For instance, when charcoal is burnt to ash, the charcoal is not really destroyed. it combines with the oxygen present in the air and becomes carbondioxide.

According to the Jain Dharma the number of substances present now, were existing in the past and will be in existence. in the future. There will be no reduction or increase in this. All transformations assume those forms according to their properties and potentialities; and in course of time, they get destroyed and cease to exist in that form. Dravya in its original and essential form remains indestructible.

The Organization of the Universe: The Jain philosophical thinking on creation and formation.

The Jain dharma believes that this universe is without a beginning; and without an end. The universe did not begin at any time and it will not end at any time. The Jain Dharma totally rejects the theory that God created this universe; or that he made this world. If the theory of creation is accepted as true, countless problems crop up.

If God created this universe, who created that God ? Who created the creator who created God ? Such questions keep cropping up and there is no end to it. Which came first? The hen or the egg? The night or the day? There is no end to these questions; and there are no answers too.

Who is God?: Why did God create this world or this universe? Only he can be called God who is omniscient, omnipotent, and who is devoid of desires, and passions.

If God really has these attributes, then why did he create this world which is groaning in agony and anguish? If really God is such why did he steep 87.5 percent of the people of this world in poverty, misery and want; and why did he bestow prosperity and happiness on the remaining 12.5 per cent? Why? If God has created such a world impelled by his desire or curiosity, then the question arises whether he can be called God at all if he is a thrall to passions and desires. How can we say that God is "Karunanidhi, (treasure of benevolence) if he causes sorrow to others for his pleasure? Desire is the greatest of all evils. With evil in him how can he be God ? How can divinity manifest itself in him?

"God gives happiness or sorrow to people, in proportion to their merits or demerits (good deeds or sins),,. If we believe this statement to be true, then the questions arises: Why did he give some people the will and ability to do good deeds? Why did he give them the virtuous mind that enables them to do good deeds? So that they may experience happiness and prosperity as a result of it or vice-versa, But why has he given a majority of people a wicked mind that compels them to do evil? Do they not, on account of it, suffer from poverty, hunger and want in their lives? They commit evil deeds; and he allows them to do them; and then he punishes them for them. What kind of justice is this? The judge himself allows the evil deeds to be committed; and hence the judge himself deserves punishment.

Instead of getting caught in the meshes of such questions, it would be better to consider the universe as being devoid of any beginning. We have to think that after all some truth is without a beginning; then why is it not possible to consider that the universe itself is without a beginning ? This belief at least frees us from the conflict of problems that defy any solution.

How can it be said that "In this universe all creatures are God's images and that they are his shadows?" This cannot be true because when God is free from all bondages all atmas-souls also should be free. If God is in bliss and if he is free from misery, then all creatures who are his images or sparks must be in bliss. There should be no need for any one to trudge upon the hard path of salvation. All should be like God. But this is not the case.

Every creature living in this universe possesses a separate and independent existence and individuality. Every one's soul is independent. Everyone has a separate existence. Every creature has to put forth its own efforts to get deliverance from the bondage of Karmas and to attain salvation because Dharma is essentially a personal affair. It is an attainment to be made by one's own self. The bondage and deliverance of each individual belong to himself or herself. The experience of happiness or sorrow belongs to each individual and it is his own. Therefore, the Jain Dharma says:

> *This universe has not been created by God; but has been revealed by him. He does not create this universe.. he does not rule it; does not govern it; and does not direct it and "the creation is such". God surely reveals to us the real nature and form of the universe. Seeing with his divine eyes, he reveals the essential and real form of the universe. The Tirthankar Bhagwant has said: The world is without a beginning. It has no beginning. It has no end. This creation is boundless being devoid of a beginning and an end. But it is present in the flow and flux of time. The universe sometimes grows small. Creation and destruction; production and disposal are always going on. Behind this eternal process there does not exist anyone's planning or organization. The whole universe is a self-regulated one. But in this organization, Karma plays an important role. In this process the effect of Karma is emphatically evident*

Jainism is properly the name of one of the religious traditions that have their origin in the Indian subcontinent. According to its own traditions, the teachings of Jainism are eternal, and hence have no founder; however, the Jainism of this age can be traced back to Mahavira, a teacher of the sixth century BCE, a contemporary of the Buddha. Like those of the Buddha, Mahavira's doctrines were formulated as a reaction to and rejection of the Brahmanism (religion based on the Hindu scriptures, the Vedas and Upanisads) then taking shape.

The brahmans taught the division of society into rigidly delineated castes, and a doctrine of reincarnation guided by karma, or merit brought about by the moral qualities of actions. Their schools of thought, since they respected the authority of the Vedas and Upanisads, were known as orthodox *darsanas* (which means literally, 'views'). Jainism and Buddhism, along with a school of materialists called *Carvaka*, were regarded as the unorthodox *darsanas*, because they taught that the Vedas and Upanisads, and hence the brahman caste, had no authority.

There are from parts of this article:

1. *Metaphysics:* According to Jain thought, the basic constituents of reality are souls (*jiva*), matter (*pudgala*), motion (*dharma*), rest (*adharma*), space (*akasa*), and time (*kala*). Space is understood to be infinite in all directions but not all of space is inhabitable. A finite region of space, usually described as taking the shape of a standing man with arms akimbo, is the only region of space that can contain anything. This is so because it is the only region of space that is pervaded with *dharma*, the principle of motion (*adharma* is not simply the absence of *dharma*, but rather a principle that causes objects to stop moving). The physical world resides in the narrow part of the middle of inhabitable space. The rest of the inhabitable universe may contain gods or other spirits. While Jainism is dualistic—that is, matter and souls are thought to be entirely different types of substance—it is frequently said to be atheistic. What is denied is a creator god above all. The universe is eternal, matter and souls being equally uncreated. The universe contains gods who may be worshipped for various reasons, but there is no being outside it exercising control over it. The gods and other

superhuman beings are all just as subject to karma and rebirth as human beings are. By their actions, souls accumulate karma, which is understood to be a kind of matter, and that accumulation draws them back into a body after death. Hence, all souls have undergone an infinite number of previous lives, and—with the exception of those who win release from the bondage of karma—will continue to reincarnate, each new life determined by the kind and amount of karma accumulated. Release is achieved by purging the soul of all karma, good and bad.Every living thing has a soul, so every living thing can be harmed or helped. For purposes of assessing the worth of actions (see Ethics, below), living things are classified in a hierarchy according to the kinds of senses they have; the more senses a being has, the more ways it can be harmed or helped. Plants, various one-celled animals, and 'elemental' beings (beings made of one of the four elements—earth, air, fire, or water) have only one sense, the sense of touch. Worms and many insects have the senses of touch and taste. Other insects, like ants and lice, have those two senses plus the sense of smell. Flies and bees, along with other higher insects, also have sight. Human beings, along with birds, fish, and most terrestrial animals, have all five senses. This complete set of senses (plus, according to some Jain thinkers, a separate faculty of consciousness) makes all kinds of knowledge available to human beings, including knowledge of the human condition and the need for liberation from rebirth.

2. *Epistemology and Logic:* Underlying Jain epistemology is the idea that reality is multifaceted (*anekanta*, or 'non-one-sided'), such that no one view can capture it in its entirety; that is, no single statement or set of statements captures the complete truth about the objects they describe. This insight, illustrated by the famous story of the blind men trying to describe an elephant, grounds both a kind of fallibilism in epistemology and a sevenfold classification of statements in logic.Every

school of Indian thought includes some judgment about the valid sources of knowledge (*pramanas*). While their lists of *pramanas* differ, they share a concern to capture the common-sense view; no Indian school is skeptical. The Jain list of *pramanas* includes sense perception, valid testimony (including scriptures), extra-sensory perception, telepathy, and *kevala*, the state of omniscience of a perfected soul. Notably absent from the list is inference, which most other Indian schools include, but Jain discussion of the *pramanas* seem to indicate that inference is included by implication in the *pramana* that provides the premises for inference. That is, inference from things learned by the senses is itself knowledge gained from the senses; inference from knowledge gained by testimony is itself knowledge gained by testimony, etc. Later Jain thinkers would add inference as a separate category, along with memory and *tarka*, the faculty by which we recognize logical relations.Since reality is multi-faceted, none of the *pramanas* gives absolute or perfect knowledge (except *kevala*, which is enjoyed only by the perfected soul, and cannot be expressed in language). As a result, any item of knowledge gained is only tentative and provisional. This is expressed in Jain philosophy in the doctrine of *naya*, or partial predication (sometimes called the doctrine of perspectives or viewpoints). According to this doctrine, any judgment is true only from the viewpoint or perspective of the judge, and ought to be so expressed. Given the multifaceted nature of reality, no one should take his or her own judgments as the final truth about the matter, excluding all other judgments. This insight generates a sevenfold classification of predications. However it is translated, it is intended to mark respect for the multifaceted nature of reality by showing a lack of conclusive certainty.Early Jain philosophical works (especially the *Tattvartha Sutra*) indicate that for any object and any predicate, all seven of these predications are true. That is to say, for every object a and every predicate F, there is some circumstance in which, or perspective from which, it is correct to make claims of each of these forms. These seven categories of predication are not to be understood as seven truth-values, since they are all seven thought to be true. Historically, this view has been criticized (by Sankara, among others) on the obvious ground of inconsistency. While both a proposition and its negation may well be assertible, it seems that the conjunction, being a contradiction, can never be even assertible, never mind true, and so the third and seventh forms of predication are never possible. This is precisely the kind of consideration that leads some commentators to understand the 'syat' operator to mean 'from a perspective.' Since it may well be that from one perspective, a is F, and from another, a is not-F, then one and the same person can appreciate those facts and assert them both together. Given the multifaceted nature of the real, every object is in one way F, and in another way not-F. An appreciation of the complexity of the real also can lead one to see that objects are, as they are in themselves, indescribable (as no description can capture their entirety). This yields the fourth form of predication, which can then be combined with other insights to yield the last three forms.Perhaps the deepest problem with this doctrine is one that troubles all forms of skepticism and fallibilism to one degree or another; it seems to be self-defeating. After all, if reality is multifaceted, and that keeps us from making absolute judgments (since my judgment and its negation will both be equally true), the doctrines that underlie Jain epistemology are themselves equally tentative. For example, take the doctrine of *anekantevada*. According to that doctrine, reality is so complex that any claim about it will necessarily fall short of complete accuracy. The doctrine itself must then fall short of complete accuracy. Therefore, we should say, "Perhaps (or "from a perspective") reality is multifaceted." At the same time, we have to grant the propriety, in some circumstances, of saying, "Perhaps reality is not multifaceted." Jain epistemology gains assertibility for its own doctrine, but at the cost of being unable to deny contradictory doctrines. What begins as a laudable fallibilism ends as an untenable relativism.

3. *Ethics:* Given that the proper goal for a Jain is release from death and rebirth, and rebirth is caused by the accumulation of karma, all Jain ethics aims at purging karma that has been accumulated, and ceasing to accumulate new karma. Like Buddhists and Hindus, Jains believe that good karma leads to better circumstances in the next life, and bad karma to worse. However, since they conceive karma to be a material substance that draws the soul back into the body, all karma, both good and bad, leads to rebirth in the body. No karma can help a person achieve liberation from rebirth. Karma comes in different kinds, according to the kind of actions and intentions that attract it. In particular, it comes from four basic sources: (1) attachment to worldly things, (2) the passions, such as anger, greed, fear, pride, etc., (3) sensual enjoyment, and (4) ignorance, or false belief. Only the first three have a directly ethical or moral upshot, since ignorance is cured by knowledge, not by moral action.The moral life, then, is in part the life devoted to breaking attachments to the world, including attachments to sensual enjoyment. Hence, the moral ideal in Jainism is an ascetic ideal. Monks (who, as in Buddhism, live by stricter rules than laymen) are constrained by five cardinal rules, the "five vows": (1) *ahimsa*, frequently translated "non-violence," or "non-harming," *satya*, or truthfulness, *asteya*, not taking anything that is not given, *brahmacharya*, chastity, and *aparigraha*, detachment. This list differs from the rules binding on Buddhists only in that Buddhism requires abstention from intoxicants, and has no separate rule against attachment to the things of the world. The cardinal rule of interaction with other *jivas* is the rule of *ahimsa.* This is because harming other *jivas* is caused by either

passions like anger, or ignorance of their nature as living beings. Consequently, Jains are required to be vegetarians. According to the earliest Jain documents, plants both are and contain living beings, although one-sensed beings, so even a vegetarian life does harm. This is why the ideal way to end one's life, for a Jain, is to sit motionless and starve to death. Mahavira himself, and other great Jain saints, are said to have died this way. That is the only way to be sure you are doing no harm to any living being.While it may seem that this code of behaviour is not really moral, since it is aimed at a specific reward for the agent—and is therefore entirely self-interested—it should be noted that the same can be said of any religion-based moral code. Furthermore, like the Hindus and Buddhists, Jains believe that the only reason that personal advantage accrues to moral behaviour is that the very structure of the universe, in the form of the law of karma, makes it so.

Jainism believes that universe and all its substances or entities are eternal. It has no beginning or end with respect to time. There is no need of some one to create or manage the affairs of the universe. Universe in run own its own accord by its own cosmic laws. Hence Jainism does not believe in God as a creator, survivor, and destroyer of the universe.

However Jainism does believe in God. When a living being destroys all his karmas, he possesses perfect knowledge, vision, power, and bliss. He becomes omniscient and omnipotent. This living being is a God of Jain religion. Hence Jains do not believe in one God. Gods in Jain religion are innumerable and the number is continuously increasing as more living beings attain liberation. Every living being has a potential to become God of the Jain religion.

A person who destroys all eight types of karmas is called Siddha. A person who destroys only four ghati karmas is called Arihanta (Tirthankara, Jina etc). Both Arihantas and Siddhas are classified as Gods in Jainism.

Now the main questions that can be arise in ones mind could be :

- What is Navakar Mantra?
- Who are Arihants?
- Who are Siddhas?
- Who are Tirthankars?
- Who are Jina?
- Who are Acharyas?
- Who are Upadhyayas?
- Who are Sadhus?
- Who are Sadhvis?
- Who are Jinas?
- What is the role of Passions in our life? What are different karmas? Why do pay homage to Arihantas first?

Namo Arihantanam : I bow down to Arihanta

Namo Siddhanam : I bow down to Siddha

Namo Ayriyanam : I bow down to Acharya

Namo Uvajjhayanam : I bow down to Upadhyaya

Namo Loe Savva-sahunam : I bow down to Sadhu

Eso Panch Namokaro : These five bowing downs

Savva-pavappanasano : Destroy all the sins

Manglananch Savvesim : Amongst all that is auspicious

Padhamam Havei Mangalam : *This Navkar Mantra is the foremost*

Now the main questions that can be arise in ones mind could be :

- What is Navakar Mantra?
- Who are Arihants?
- Who are Siddhas?
- Who are Tirthankars?
- Who are Jina?
- Who are Acharyas?

- Who are Upadhyayas?
- Who are Sadhus?
- Who are Sadhvis?
- Who are Jinas?
- What is the role of Passions in our life? What are different karmas? Why do pay homage to Arihantas first?

The Navkar Mantra is the most fundamental mantra in Jainism and can be recited at any time of the day. While reciting the Navkar Mantra, the aspirant bows with respect to Arihantas, Siddhas, Acharyas, Upadhyayas, Sadhus, and Sadhvis. The mantra enables us to worship the virtues of all the supreme spiritual people instead of just worshipping one particular person. For this reason, the Navkar Mantra does not mention the names of any Tirthankaras, Siddhas, Acharyas, Upadhyayas, Sadhus, or Sadhvis.

At the time of recitation, we remember their virtues and try to emulate them. In this mantra we bow down to these supreme spiritual personalities, and therefore, it is also called Namaskar or Namokar Mantra. The Navkar Mantra contains the essence of Jainism. It points out that if we want to be truly liberated, we have to give up worldly life (samsar). The first stage of renunciation is to become a monk (sadhu) or nun (sadhvi). While progressing on a spiritual path, some may be designated as Upadhyayas or Acharya. The ultimate aim is to attain omniscience, becoming an Arihanta, which leads us to liberation, the becoming a Siddha.

Arihantas: The term Arihanta is made up of Ari, meaning enemies, and hant, meaning destroyer. Consequently, Arihanta means destroyer of enemies. In this case the term enemies refers to passions such as anger, greed, ego, and deceit which are internal enemies, because they defile the true nature of the soul. A soul can only reach the state of Arihanta by overcoming all its inner enemies. Once a soul has shed all of its four defiling (ghati) karmas namely Jnanavarniya (Knowledge obscuring) Karma, Darshanavarniya (Perception obscuring) karma, Mohniya (Deluding) Karma and Antaraya (Obstructive) Karma, it becomes an Arihanta and attains perfect knowledge (Kevaljnana), perfect perception (Kevaldarshana), and infinite power (Ananta Virya) and it becomes a passionless (vitragi). Arihantas are divided into two categories: Tirthankar and Ordinary. Arihantas who have attained Tirthankar Nam Karma become Tirthankaras while the rest of them become Ordinary Arihants.

There are twenty-four Tirthankaras during every half time cycle. These Tirthankaras reinstate the Jain Sangh (four-fold Jain Order) consisting of Sadhus (monks), Sadhvis (nuns), Shravaks (male householders), and Shravikas (female householders). The first Tirthankar (Arihanta) of this time period was Lord Rushabhdev, and the twenty-fourth and last Tirthankar was Lord Mahavira, who was lived from 599 B.C. to 527 B.C. Tirthankaras are also called Jina (conqueror of inner passions) from which the term Jain, follower of a Jina, is derived. At the time of Arihanta's nirvan (death), the remaining four non-defiling (aghati) karmas such as Nam (Physique determining) Karma, Gotra (Status determining) Karma, Vedniya (Feeling producing) Karma and Ayushya (Age span determining) Karma, are destroyed. Ordinary Arihants are those souls who attain salvation, but do not possess Tirthankar Nama Karma and hence, do not establish the Jain Order.

After attaining salvation they are called Siddhas. Since Siddhas have attained ultimate liberation, we do not have access to them. However, Arihantas offer us spiritual guidance during their lifetime. In order to show our special reverence for their teachings, we bow to them first, hence the first verse of the Navkar Mantra. Currently, as per scriptures except at Mahavideh kshetra, there are no Arihantas. The last Arihant was Jambuswami. According to the Agams (Jain scriptures) there will be no more Arihantas during the remaining period of the current half-time cycle.

Siddhas: Siddhas are liberated souls. They have reached the highest state, salvation, and have attained Moksha. They have eradicated all their karmas, and therefore do not accumulate any more new karmas, thus freeing themselves forever from the cycle of birth and death (Akshaya Sthiti). This state of freedom is called Moksha. They are experiencing ultimate, unobstructed bliss (Aksha Sukh) and are not subjected to any kind of suffering.

They possess perfect and total knowledge (Anatjnan, Kevaljnana, omniscience) and perception (Anat Darshan, Kevaldarshana, omniperception), that means they know and perceive everything in total that is happening now, that has happened in the past, and that which will happen in the future all at the same time and they also possess infinite vigor (Anant-Virya). They have no desires and are completely detached thus making them immune from any sense of craving or aversion (Anant Charitra, Vitragatva). Despite the fact that all Siddhas retain a unique identity, they are equal (Aguru-laghutva) and formless(Arupitva).

Acharyas: The message of Jina, Lord Mahavira the last Tirthankara, is carried by the Acharya, our spiritual leaders. The responsibility of the spiritual welfare of the entire Jain Sangh rests on the shoulders of the Acharyas. Before reaching this state, one has to do an in-depth study and have a thorough mastery of the Jain Agams. In addition to acquiring a high level of spiritual excellence, they also have the ability to lead the monastic communion. They should also know the various languages of the country and have acquired a sound knowledge of other philosophies, ideologies, and religions of the region and the world.

Upadhyayas: This title is given to those Sadhus who have acquired a special knowledge of the Agams (Jain scriptures) and philosophical systems. They teach Jain scriptures to deserving aspirants, including sadhus and sadhvis.

Sadhus (Muni): A male person who renounces the worldly life is called a monk or sadhu, and a female is called a nun or sadhvi. When householders become detached from the worldly aspects of life and aspire for spiritual uplift, they renounce their worldly lives and become Sadhus or Sadhvis, by accepting Deeksha. Before such initiation, they must stay with Sadhus or Sadhvis for a period of time to understand religious studies and to observe the code of conduct for renounced life. When they feel confident, they request an Acharya to initiate them into the renounced order. If the Acharya feels that they have the desire and capability to face the rigors of renounced life, then he gives them Deeksha. At the time of Deeksha, the newly initiated sadhu or sadhvi adopts five major vows:

1. Observance of Ahimsa (non-violence)-not to commit any type of violence (Savvao Panaivayao Virman Vrat)
2. Observance of Satya (truth)-not to indulge in any type of lie or falsehood (Savvao Musavayao Virman Vrat)
3. Observance of Asteya (non-stealing)-not to take anything unless it is given by the owner (Savvao Aadinnadanao Virman Vrat)
4. Observance of Brahamcharya (celibacy)-not to indulge in any sensual pleasure (Savvao Mehunao Virman Vrat)
5. Observance of Aparigraha (non-possessiveness)-not to acquire more than what is needed to maintain day to day life (Savvao Pariggrahao Virman Vrat).

Some of the special things they observe are they do not accept the food cooked for them. They do not eat before sunrise or after sunset. They drink only boiled water. They walk bare feet. They do not stay in one place for a longer time. They do not touch any person of opposite sex. They do not get involved in social affairs there by meaning they are not a social workers. All nuns wear white clothes. They offer spiritual guidance to us. Their goal to become is to be liberated from this worldly life and that is why their activities are directed towards uplift of their souls to Paramatman (the state of liberation). Self-discipline and purity is the main part of their daily lives.

Nav Padhrtha (Elements): Jain philosophy can be described in various ways, but the most acceptable tradition is to describe it in terms of the Nav Padhrtha (Elements) or nine fundamentals. They are:

1. Jiva (soul)
2. Ajiva (non-living matter)
3. Punya (results of good deeds)
4. Pap (results of bad deeds)

5. Asrava (influx of karmas)
6. Samvar (stoppage of karmas)
7. Bandh (bondage of karmas)
8. Nirjara (eradication of karmas)
9. Moksha (liberation)

Now, let us use a simple analogy to illustrate these Padhrtha (Elements). There lived a family in a farm house. They were enjoying the fresh cool breeze coming through the open doors and windows.

The weather suddenly changed, and a terrible dust storm set in. Realizing it was a bad storm, they got up to close the doors and windows. By the time they could close all the doors and windows, lots of dust had entered the house. After closing the doors and the windows, they started clearing the dust that had come in to make the house clean.

We can interpret this simple illustration in terms of Nav-Padhrtha (Elements) as follows:

1. Jivas are represented by the people.
2. Ajiva is represented by the house.
3. Punya is represented by enjoyment resulting from the nice cool breeze.
4. Pap is represented by discomfort resulting from the sand storm, which brought dust into the house.
5. Asrava is represented by the influx of dust through the doors and windows of the house which is similar to the influx of karman particles to the soul.
6. Bandh is represented by the accumulation of dust in the house, which is similar to bondage of karman particles to the soul.
7. Samvar is represented by the closing of the doors and windows to stop the dust from coming into the house, which is similar to the stoppage of influx of karman particles to the soul.
8. Nirjara is represented by the cleaning up of accumulated dust from the house, which is similar to shedding accumulated karmic particles from the soul.

9. Moksha is represented by the cleaned house, which is similar to the shedding off all karmic particles from the soul.

Jiva: All living beings are called Jivas. jivas have a consciousness known as the soul, which is also called the atma (soul-chetan). The soul and body are two different entities. The soul cannot be reproduced. It is described as a sort of energy which is indestructible, invisible, and shapeless. Jainism divides jivas into five categories ranging from one-sensed beings to five-sensed beings. The body is merely a home for the soul. At the time of death, the soul leaves the body to occupy a new one. Tirthankaras have said that the soul has an infinite capacity to know and perceive. This capacity of the soul is not experienced in its present state because of accumulated karmas.

Ajiva: Anything that does not have an soul is called ajiva. Ajiva does not have consciousness. Jainism divides ajiva in five broad categories: dharmastikay (medium of motion), adharmastikay (medium of rest), akashastikay (space), pudgalastikay (matter), and kala (time).

Punya: By undertaking wholesome activities, we acquire punya or good karmas. Some such activities are providing food or other items to the needy people, doing charity work, propagating religion, etc. When punya matures, it brings forth comfort and happiness.

Pap: By undertaking bad activities, we incur pap or bad karmas. Some such activities are being cruel or violent, showing disrespect to parents or teachers, being angry or greedy and showing arrogance or indulging in deceit. When pap matures, it brings forth suffering, misery, and unhappiness.

Asrava: The influx of karman particles to the soul is known as asrava. It is caused by wrong belief, vowlessness (observing no vows), passions, negligence, and psychophysical activities. Such and influx of karmas is facilitated by mental, verbal, or physical activities.

Samvar: This is the process by which the influx of karman particles is stopped. This is achieved by observing samiti (carefulness), gupti (control), ten fold yati-dharma (Monkshood), and contemplating on the twelve bhavanas (mental reflections), and parishaha (suffering).

Bandh: This refers to the actual binding of karman particles to the soul. Bandh occurs when we react any situation with a sense of attachment or aversion.

Nirjara: The process through which we shed karmas is called nirjara. Karmas can be shed either by passivity or active efforts. When we passively wait for karmas to mature and to give their results in due time, it is called Akam Nirjara. On the other side, if we put active efforts for karmas to mature earlier than supposed to be, it is called Sakam Nirjara. Sakam Nirjara can be achieved by performing penance, regretting, asking for forgiveness for the discomfort we might have caused to someone, or meditation, etc.

Moksha: If we rid ourselves of all karmas, we will attain moksha or liberation.

Leshyas in Jainism, there is a great deal of importance given to Leshya. Leshya refers to the state of Mind. Our activities reflect our state of mind. The following illustration shows how our activities vary with the state of our mind.Once there were six friends who were going on a trip, but along the way got lost in a forest. Soon they became hungry and thirsty. They searched for food for sometime and finally found a fruit tree.As they ran to the tree, the first man said, "Let's cut the tree down and get the fruits." The second one said, "Don't cut the whole tree down, cut off a big branch instead." The third friend said, "Why do we need a big branch? A small branch has enough fruit." The fourth one said, "We do not need to cut any branches, just climb up and take the bunches of fruit."

The fifth man said, "Why pick that many fruits and waste them, instead just pick the fruits that you want to eat." The sixth friend said quietly, "There are plenty of good fruits on the ground, so let's just eat them first."You can see that the states of minds of these six friends cause a range of thoughts which begin with the destruction of a tree and end with picking up what is on the ground. The six friends' minds represent six types of leshyas.The first friend's state of mind represents krishna (black) leshya. The second friend's state of mind represents neel (blue) leshya. The third friend's state of mind represents kapot (brown) leshya.The fourth friend's state of mind represents tejo (red) leshya.The fifth friend's state of mind represents padma (yellow) leshya.The sixth friend's state of mind represents shukla (white) leshya.The first leshya is the worst and the sixth leshya is the best. The first three leshyas lead the soul to ruin and the last three lead the soul to spiritual prosperity. Therefore, people should strive for the white leshya and not the red leshya.

It is not that a person's state of mind remains the same at all times and therefore, it can be changed for the better or for the worst. The popular story of King Prasanraj, who lived during Lord Mahavira's time, illustrates the effects of leshays. One day King Shrenik was on his way to pay homage to Lord Mahavira and on the way he saw a sage who was meditating and illuminating very brightly. He bowed down to the sage and continued on his way to pay homage to Lord Mahavira. After reaching Lord Mahavira, King Shrenik asked the Lord, "Oh Lord, I saw a very illuminated sage who was engaged in meditation on the way here. Supposing he died at that moment, what state would he have acquired?"

The Lord replied, "He would have been hurled down to the seventh hell-region." The king was much astonished to hear this reply from the Lord. He thought," Why would such a sage go to hell? Perhaps the Lord might have misunderstood me." He asked the Lord again, "Oh Lord, if the sage leaves this body just now, then what state will he attain?" The Lord replied, "He will be an angel in Sarvarthasidha, a heavenly region." The king was much surprised at this reply, too. He thought, "The Lord first said he would attain the seventh hell and now he says that the sage would be an angel."

The king was perplexed. At that very moment drums began sounding in the sky and voices of 'victory' were proclaimed. The king asked the Lord, "What is the cause of these sounds?" The Lord said, "Oh, king, the sage about whom you were inquiring has acquired omniscience and so the angels are beating the drums and proclaiming `Victory'." The king was extremely surprised by these answers and therefore, he requested the Lord to explain them. Thereupon the Lord said, "Oh king, right before you approached the sage, two soldiers leading your procession diverted the mind of the sage by their conversation. His mind began thinking about his kingdom where his ministers turned out to be untrustworthy and would eventually kill his son. He was inflamed with rage and he lost his mental equanimity.

Therefore, he started mentally fighting against his ministers. He very violently discharged his weapons one after the other against his ministers. Soon his weapons were exhausted and his foes were not destroyed. So, he thought of throwing his steel helmet against them in order to destroy them. You bowed down to him at that time when he was reflecting on such thoughts. I, Therefore, replied that he would attain the 7th hell. Thereafter, when the sage placed his hand on his head for the steel helmet, he realized that he was not King Prasanraj but that he was a sage. His anger calmed down immediately. He remembered that he has been initiated into the vow of equanimity and of nonviolence to any living being mentally, verbally, and physically. He deeply regretted and repented for the breach of his vow and indulgence in acute anger. He further thought that he ought to have maintained love for all creatures of the world, ought to have no malice for the ministers, and no attachment for his son. He severely condemned his mental act. He despised it and withdrew himself from such a feat of anger and malice. Oh king, when he thought this way, you asked me the next question and I replied that he would be born in the Savartha Siddha (heaven) as an angel. Even thereafter, he continued the purification of his mental reflections and gradually he reached the stage of 'Kshapaka', where he annihilated all of his ghati karmas, and attained omniscience." King Shrenik's doubts were resolved and he learned how mental reflections can fluctuate. He also learned that not only can physical acts or verbal abuses have such devastating effects, but so can mental acts. We too must learn from this episode. Let us understand the characteristic of personalities and the outcomes resulting different leshyas.

1. *Krishna (Black) Leshya:* People in this state of mind do not show no compassion or mercy. Everyone is afraid of them as their anger turns into violence. They always burn with jealousy and have ill-will for everyone. They are filled with animosity and malice and do not believe in religion. This state of mind is the worst and most dangerous. If anyone dies in this state of mind, they will go to hell.
2. *Neel (Blue) Leshya:* People in this state of mind are proud, haughty, and lazy. They are unreliable and other people avoid their company. They are cheaters, cowards, and hypocrites. Such people also avoid the religious lectures. If anyone dies in this state of mind, they will be reborn as a plant.
3. *Kapot (Brown) Leshya:* People in this state of mind always remain sad and gloomy. They find faults in others and are vindictive. They boast about themselves, become excited over small matters, and lack mental balance. If anyone dies in this state of mind, they will be reborn as a bird or an animal.
4. *Tejo (Red) Leshya:* People in this state of mind are very careful about their actions and discriminate between good and evil. They know the difference between what is right and what is wrong. They are kind, benevolent, and religious and lead a harmonious life. If anyone dies in this state of mind, they will be reborn as a human being.
5. *Padma (Yellow) Leshya:* People in this state of mind are kind and benevolent and forgive everyone, even their enemies. They observe some austerities and are vigilant in keeping their vows till their last breath. They remain unaffected by joys and sorrows. If anyone dies in this leshya, they will be reborn in heaven as a celestial being.
6. *Shukla (White) Leshya:* People in this state of mind strictly observe the principles of nonviolence, truth, non-stealing, celibacy, and non-attachment. They are

trustworthy, treat every soul as if it was their own soul, and do not have ill feelings even for their enemies.

They remain calm even if someone abuses them. If anyone dies in this state of mind, they will be reborn as a human being or an angel. People who perfect this state of mind have no attachment or hatred and treat everyone alike. They do not become happy or sad. Their state of mind is the purest. If anyone dies in this perfected state of mind, they will be liberated from the cycle of birth and death.

Bhavanas: Bhavanas means reflections or contemplation. We are always thinking of something or another which may be near us or could even be as far away as the other end of the world. This thinking may cause happiness, sorrow, anger, greed, jealousy, and egoism, etc. And, depending upon how we react we attract various types of karmas to our souls. No prudent person would like to attract bad karma and at certain stages any karma at all. But it is not as easy as turning off a light switch. To minimize the bondage of karmas, the Jina have taught us to observe sixteen Bhavanas. Of them the following four can easily be practiced by householders while the rest may be observed once we become more serious about spiritual uplift.

1. Maitri Bhavana (Thinking of being a Friend)
2. Pramod Bhavana (Thinking of Appreciation)
3. Karuna Bhavana (Thinking of Compassion)
4. Madhyastha Bhavana (Think to staying neutral or uninvolved)

Let us understand how these bhavanas play important roles in our lives, how they can keep us away from bad karmas, and ultimately how they can improve our overall chances to uplift our souls.

Maitri Bhavana (Thinking of a Friendship): Lord Mahavira said that we must be a friend to all living beings. The feeling of friendship brings love and respect for others. It also initiates a feeling of brotherhood among all and in turn leaves no room for harm, deceit or quarrelsomeness with anybody. If we contemplate on Maitri Bhavana our thoughts, words, and actions will not be harsh and we will not hurt anybody. On the contrary, we will support and protect everybody. Friendship will lead us to be tolerant, forgiving and caring for one another. It can be seen that if we develop a friendship with all living beings, we will avoid bad karma.

Pramod Bhavana (Thinking of Appreciation): In this bhavana we admire the success of our friends, spiritual leaders and Arihants. One of the most destructive forces in our lives is jealousy. However, friendship combined by admiration, destroys jealousy. As jealousy subsides, negative impulses are turned into positive ones, and in time you will be at peace.

Karuna Bhavna (Thinking of Compassion): Instead of succeeding, many of our friends may be getting into trouble for things they should not be doing and even those who are successful may be accumulating vices such as greed and ego. If such is the case then, they are not on the right path and may be weak, helpless, and in distress. At a time like this you should contemplate on the karuna Bhavana and show compassion towards them instead of disgusted towards them or hating them. Show them a right path with patience, tolerance, and forgiveness and offer them needed support. In this way not only will you be avoiding bad karma, but so will your friends.

Madhyastha Bhavana (Think of staying neutral or uninvolved) : Life appears to be nothing but involvements. Sometimes life works out favorable and sometimes it does not. So instead of getting disappointed, angry, or more involved, you should contemplate on

Madhyastha Bhavana which leads to feeling that 'I did my best to resolve the situation.' If someone does not want to understand, then leave them alone without getting further involved. You should simply hope that one of these days they may understand and change. By observing Madhyastha Bhavana you will remain in equanimity instead of provoking turmoil in your mind. When your mind stays neutral and uninvolved it keeps karma away.

In short, you can avoid influx of bad karma and stay in peace while living in worldly life by developing a friendship towards all living beings, admiring their success, holding their hands when they are in distress, and leaving them alone at times when they do not understand what is right or wrong. So until it becomes the natural way of life to observe the above bhavanas, one should contemplate on them as many times as needed. If there is a goal, then there is an achievement!

(There are Twelve More Bhavnas which are not discussed here at this point)

Charitra: To achieve Moksha we need the right knowledge, right faith, and right conduct known as three jewels of Jainism. Regarding right conduct, we must achieve control over our inner desires and reach a stage where there is no attachment or hatred. It is not as easy as it sounds, for we will have to form new habits and discard old habits.

In order to change our habits, we take vows which help us to restrict what we do and eventually live naturally restrained. Although different people take different vows and despite the outer differences in the observance of these vows, the goal of all is to attain right conduct. As we find in the scriptures, there are different vows for monks and nuns, and for male and female house holders.

When one makes a resolution to restrain willfully from something with full understanding and faith then that resolution becomes a vow. The scriptures have divided these vows into two main groups:

1. Maha Vrats (Major Vows or Absolute Vows)
2. Anu Vrats (Minor Vows or Relative Vows)

Maha Vrats

1. *Savvao Panaivayao Virman Vrat (Absolute Non-Violence Vow)*

   In this vow, nonviolence is observed mentally, verbally, and physically in total including for daily needs.

2. Savvao Musavayao Virman Vrat (Absolute Truthfulness Vow)

   In this vow, lying is given up mentally, verbally, and physically in total. Only the truth is spoken.

3. Savvao Aadinnadanao Virman Vrat (Absolute Non-Stealing Vow

   In this vow, stealing is given up mentally, verbally, and physically in total. Nothing is taken unless it is offered by the owner, and meets the restrictions of the monkshood life.

4. Savvao Mehunao Virman Vrat (Absolute Celibacy Vow)

   Celibacy is observed mentally, verbally, and physically in total. Upon taking this vow, one can not even touch or think about a member of the opposite sex.

5. Savvao Pariggrahao Virman Vrat (Absolute Non-Possessiveness Vow)

   Greed and material possessions are given up mentally, verbally, and physically in total. Only items which are needed on a daily basis are kept.

Anu Vrats

1. Sthool Panaivayao Virman Vrat (Limited Non-Violence Vow)

   Being a house holder, you must carry out house hold and community activities. Thus, it is not possible to totally avoid violence, especially against one sensed living beings. However, you should take the utmost care to observe non-violence.

2. Sthool Musavayao Virman Vrat (Limited Truthfulness Vow)

   You should take the utmost care in not telling lies. Another words you should only speak the truth.

3. Sthool Aadinnadanao Virman Vrat (Limited Non-Stealing Vow)

   You should take the utmost care not to steal or borrow the belongings of others without their permission.

4. Sthool Mehunao Virman Vrat (Limited Celibacy Vow)

   You should be satisfied with your spouse only and should avoid sexual contact with others.

5. Sthool Pariggahao Virman Vrat (Limited Non-Possessiveness Vow) Your earning for material

possessions, and greed for collection should be as limited as possible.

Guna Vrats (Supporting Vows)

The following vows support the above five vows so that they can be carried out more effectively.

6. Disha Pariman Vrat (Geographic Limitation Vow)

   Since it is necessary to move only so much to satisfy your needs, you should limit your desire to perform activities within a needed radius of area. Therefore, you should set limits within the scope of travel or business, etc.

7. Bhogopbhog Virman (Consumption and Occupation Limitation Vow)

   You should set limits on the consumption of various items, despite whether they are used once or repeatedly, to avoid incoming sins. Also, you should limit yourself to the kind of occupation you select because some occupations have more exposure to sins than others.

8. Anartha Dand Virati Vrat (Purposeless Violence Limitation Vow)

   We sometimes carry out activities for fun or to tease our friends and foes. However, we should realize that all such activities attract karmas. This vow reminds us not to build unnecessary karma.

Shiksha Vrat (Training Vows): The above eight vows tell us how we can limit our actions in order to reduce the inflow of karma. The next four vows train us to stop karma from coming in, and to get rid of existing karmas.

9. *Samayik Vrat (Self-control and Equanimity Vow)*

   Since we live a social life, it may not be possible to spend much time in shedding karmas, but we should spend at least forty-eight minutes a day in such efforts. During these forty-eight minutes, we give up all familial, social, economical, and cultural responsibilities, and we control ourselves mentally, verbally, and physically. During this time we should concentrate on religious readings or meditation, which will help prevent new karma from coming in as well as help get rid of our old karma. This 48 minutes will also give us a glimpse as to how easy or difficult it may be to control ourselves. Once we become accustomed to doing this, we may extend the self-control time to a period of more than 48 minutes.

10. Dishavakashik Vrat (Day to Day Additional Limitation Vow)

    At the time of waking up, we take this vow in order to put definite limits on our activities for that day, and thus, limit our exposure to accumulating karma. In this way we learn to live a more controlled life.

11. Poshadh Vrat (Monkshood Exposure Vow)

    This vow teaches us how to live like a monk, or a nun, for a day. By taking this vow, we give up all our household, social, economical, and cultural responsibilities for a day. We should practice this vow as often as possible. By practicing this vow, the inflow of new karma is slowed down, and existing karmas are eradicated. If this vow is easy to observe it may eventually lead to monkshood.

12. Atithi Savinbhag Vrat (Selfless Offerings to Unexpected Guests Vow)

    This vow teaches us to share. Sharing is more worthwhile when an unexpected, needy person comes to your door and you offer with an open heart. Monks and nuns are the best receptors for this type of sharing. However, while giving, you should not have any regrets or expectations in your mind. By sharing, you are developing a sense of detachment with the feeling that nothing is mine. Observing this vow will lead to liberation from worldly life.

Our thoughts while taking these vows are also very important. They should be positive and meant to destroy karmas.

There are three common obstacles to our vows:

1. Maya (Deception)

2. Niyanu (Expectations)
3. Mithyatva (Wrong Beliefs)

Maya (Deception): When you take a vow, it should be for the betterment of your soul. Vows should not be taken to show-off or receive praise. Also, they should not to be taken to deceive or they will produce inferior results.

Niyanu (Expectations): Many of us take vows in order to gain something materialistic in return. Such is not good because not only do we use up what we achieve, but we lose the main purpose of performing vows, to destroy karmas.

Instinct: Instincts play a significant role in our life. They affect our conduct as well as behaviour. They have been illustrated in Jain Canons from the karmic philosophical point of view. One of the modern disciplines of science is psychology of behaviour which emphasizes the original instinct in the form of natural tendencies and modes of living, emotions, feelings. It corroborates the karmic explanations.

First of all we should understand the concept of an instinct. There are two meanings. One is an emotion which is called an original instinct in the psychological term and the other is consciousness. Whatever we do, is influenced by the impelling force of these special types of emotions. In the other words instict (samjna) means particular type of tendency of consciousness in which conscious and subconscious minds are combined together and called emotions. Internal and external causes activate them to arise. The material cause of them is karmas. Simultaneously external environment also affects in their production.

On the basis of the causes of generation we can divide them into several types. Thanang Sutra one of the Agama describes ten types of instincts. The first eight are emotional or sentimental instincts whereas rest of the two are concerned with a special type of thought process.

Instinct of appetite	Ahara samjna
Instinct of fear	Bhaya samjna
Instinct of copulation	Maithun samjna

Instinct of possession	Parigraha samjna
Instinct of anger	Krodha samjna
Instinct of Pride	Mana samjna
Instinct of deceit	Maya samjna
Instinct of greed	Lobha samjna
Instinct of collective mind	Ogha samjna
Instinct of individual	Loka samjna

Consciousness: The conduct and behaviour are the outcome of deluding karma (Mohaniya Karma). Due to the influence of instinct man changes his habits and nature. Therefore the above stated classification may be clearly understood in the light of karmas philosophy and psychological divisions.

Outcome of deluding karmas	***Original emotions***
Fear	Fear
Anger	Anger
Disgust (jugupsa)	Hatred
Feminine inclination (striveda)	Sexuality
Masculine inclination (purusveda)	Sexuality
Common inclination (napunskaveda)	Sexuality
Pride	Superiority complex
Greed	Ownership
Indulgence (Rati)	Attachment

The similar division is given by the well-known psychologist Mac Dowell in psychological terms as under :

Original tendencies	***Original emotions***
Tendency of escapism	Fear
Tendency of struggle	Anger
Tendency of curiosity	Eagerness
Tendency of search of food	Appetite
Tendency of paternity	Affection

Tendency of disgust	Hatred
Tendency of community feeling	Community feeling
Tendency of copulation	Sex
Tendency of prejudice	Superiority
Tendency of inferiority complex	Inferiority
Tendency of Earning	Ownership
Tendency of Creation	Creativity
Tendency of begging	Pity
Tendency of humor	Pleasure

The above given divisions can be concise in three categories. They are:

In the First Group : Instinct of appetite, fear, copulation and possessiveness. These exist in every creature of the world. An instinct of appetite is found in all. On account of this creature feels hunger and searches for it. Really a big part of our life is motivated by it. A number of behaviour of a living being are activated by fear. The feeling of getting hurt creates fear. All creatures have the instinct of copulation. The fourth one is the instinct of possessiveness exists in human beings as well as in animals. They collect their requirements and want to get more and more things due to this.

The Second Group of Instincts : Anger, pride, deceit and greed. This group of instincts also exists in all living beings. But in man they are well developed and expressed.

The first group of instincts is precursor of activating these instincts. For example when a person gets hungry and if it is not satisfies he becomes angry. To earn money people use deceive methods. Once you start earning greed sets in which may become the cause of violences, corruptions, eetc. In this way it can be said that the former group of instincts activate the later or both are correlated with each other.

The Third Group of Instincts: Community feeling and individuality. The most living beings more or less desire to live in a group. Anthropologist Maxmullar also accepts this tendency. Not only in human beings but it exists in animals also. The consciousness of individuality are found in all the animates. Every creature has some sort of distinguished characteristics.

Origin of instincts: Now the question arises what is responsible of outpouring of instincts? In this respect we illustrate internal and external causes.

Instinct of Appetite Internal: due to the rising of feeling and deluding karmas. External: due to Emptiness of stomach seeing food-stuffs thinking about food constantly smelling the flavor

Instinct of Fear: Internal: deluding karmas

Instinct of Copulation: Internal: deluding karmas

External: Strengthening of muscles and organs of the body

Listening or reading the copulation connected material. Constant thinking about it.

Instinct of Possession : Internal: deluding karmas

External: To have possession already

To listen about possession

To think about possession

In this way instincts of anger, pride etc., are also outpoured by deluding karmas and activated by external situations.

Impressions of Instincts: They are not only connected with our feelings but with physical and mental functioning structure also. They influence the external as well as internal functioning of the body. Mental states and feelings are also affected by them. We can understand them as:

Internal Changes

States of the mind

Feelings of the consciousness

Psychological responses

Nervous system

Secretion of glands

Chemistry of blood

The Scientific Approach: For a long time it was believed that the brain was the source of energy and the seat of impulses of man. Significant progresses in endocrinology (branch of science dealing with endocrine system) in the recent years has established that the passions and expressions affect the endocrine system.

Endocrine system is the basis of habit formation in a person. They manifest themselves in the Nervous system and becomes behavioral pattern. Thus manifestation and behaviour are the function of the nervous system, but the birth place-the production centre of the instincts is endocrine system. The endocrine glands control through chemical regulators called hormones. They regulate the body functions.

These chemical messengers exert profound influence on the mental states, emotions and behavioral patterns of an individual. They are primary movers which generate the powerful forces, all the instincts and urges, passions and emotions in man. These, in turn generate feelings and results into action that satisfy the needs of love, hate, fear etc. are endocrines expressions. Therefore it is said that endocrine system is the seat of the instincts from the gross point of view but actually the karmas are responsible for the producing of the instincts.

How the Instincts can be Changed?: Now the most important task before us is how these instincts and urges, impulses and feelings can be pacified? In this context you can be benefited by the regular practice of special technique of preksha meditation. There are various methods available in this field. One of them is visualization of bright white colour at the centre of enlightenment, that is in the middle part of the forehead.

If a person practises it constantly, he can control his anger and increase his power of tolerance to a great extent. The another significant process of preksha meditation is perception of deep breathing which helps to pacify the intensity of the instincts. Therefore, regular practice of meditation develops balanced and an integrated personality by decreasing the intensity of instincts.

Jain Mythology: In traditional Jain mythology, the universe is a finite three dimensional structure shaped like a man standing with hands on hips, elbows turned outwards and legs held apart. The structure is divided in to three sections, the uppermost is for the heavenly beings; the humans are located in part of the middle section and different levels of hell are located in the lower section, with the severest of hell at the bottom.

The apex of the upper section is the Siddha Shila, where the Atma resides permanently after its final salvation, moksha. The Atma can accede to moksha only from a human birth. There are several different worlds in the centre section; humans can exist in some of the inner parts of this section. (There are two main worlds that are of interest to us – Bharat khsetra, the world where we currently reside and Mahavideh khsetra about which we will talk later.)

The final release from bondage is achieved when all the karmas accumulated by that Atma (during all its previous incarnations) have been completely annihilated.

The Atmas (Souls) go through infinite reincarnations wandering between the heavenly abodes, middle sections and the hellish regions–incurring new karmic bondages whilst expending previously acquired karmas. Karma is acquired each time the Atma inspires any feeling of like, dislike, anger, greed, compassion, hate. There are good and bad karmas. Each karma will bear fruit which will have to be suffered by the Atma in one of its present or future reincarnations in one of the four modalities of birth, namely: human, animal, heavenly or hellish, each incarnation being the result of the karmas incurred in the previous lives.

Periodic Time Cycles: Jain philosophy envisions an eternal universe, cycling through ascending and descending cosmic cycles of time with no beginning and no end. Each cycle has six ascending or evolutionary phases and six descending or regressive phases. Each cycle is of extremely long but finite duration. Currently we are at the start of the fifth phase of the regressive half of the Cosmic cycle which will last for a total of 21,000 years of which just over 2,500 years have passed. It is believed that the descending cycle begins with all human beings full of goodness leading extraordinarily happy lives with all their desires and wants satisfied. During the subsequent phases the level of happiness regresses until the fifth cycle when the lives are definitely unhappy and in the last cycle they are extremely unhappy. In the evolutionary or ascending cycle the phases are retraced with progressively increasing happiness.

It is at the end of the third phase and during the fourth phase of each half cycle that humans are sufficiently unhappy to start contemplating the reasons for their impermanence and sufferings. It is during this period that twenty-four Jinas are born.

Tirthankara: A Jina is born a normal human, having progressively cleansed the soul of acquired karmas in several of his past incarnations. The Jina is born with immense physical strength, and wisdom (avdhi gnan) that enables him to foresee the events which will take place in the future. During his life time as Jina, he rids the soul off the remaining karmas (except lifespan related karma) and attains total wisdom (keval gyan) and becomes omniscient.

Each Jina would re-establish an order in the society, of the monks and laity and becomes a symbol for pilgrimage and for this reason he is called 'Tirthankara'. He sets up the criteria for the teaching to the laity. On the expiry of lifespan karma he is free of all bondages and the Jina's Atma achieves moksha, liberation. It is now called 'Siddha' and ascends to rest on the Siddha Shila as only pure knowledge and total bliss. The last Jina in our world was Bhagwan Mahavira who was born just over 2600 years ago right at the end of the 4th phase.

The teachings of a Tirthankara comprises of understanding the relationship between the living beings and non-living matter, how they are bound together by the karmas and most importantly that all living beings or 'Atmas' are intrinsically equal, immortal having innate qualities of pure knowledge, perception, complete bliss, and consciousness. Each Atma is responsible for its own embodiment and reincarnations and can attain freedom from all its bondages only through its own efforts though it must rely on teachings of those who have already attained that freedom. Jains treat the Tirthankaras as their main objects of worship and as Gods who have shown the way to salvation. However, Jains do not consider the Tirthankaras as the creators.

The Building Blocks: At the beginning of the chapter, reference was made to three concepts of Jainism which makes it a unique school of thought, two of these were; 'ahimsa' the equality of all living beings and 'ankekanta' or 'anekantvad' the multi faceted view or manifold aspects which we have reviewed above. The third building block 'aprigraha' means total detachment from all possessions and desires of possession – these include material objects as well as mental concepts of belonging (eg. my family, my mother etc). The equanimous state can only be attained when there are no expectations or desires – this can only happen when the Atma is totally detached from everything and has experienced its own nature.

The Delusion: Superficially it may appear that to develop these three building blocks may not be that difficult; however, as the Atma has been conditioned over a very long period (thousands of incarnations) to believe that itself and the body it is born into are one and this delusion clouds its ability to know and understand its own nature. The delusion of being the body leads the Atma to its attachment to the body and thereafter other attachments follow almost naturally.

The Path: This delusion can only be eliminated through the guidance of a teacher who has eliminated his own delusion and experienced his Atma. The Tirthankaras have experienced the Atma, are equanimous and totally detached. Each Tirthankara would set up through his disciples an order of monks and laity through which his teachings are disseminated.

The last Tirthankara of this regressive cycle, Lord Mahavira was born just before the end of the fourth phase as Jains believe that those born in this world in the fifth phase cannot achieve enlightenment in this lifetime. Lord Mahavira, through a question and answer session with his disciples, had preached and at the same time formulated a life style based on a system of thought processes whereby the Atma firstly understands exactly what its own nature is, why it is in that particular body at the time in question and then tries to ascertain what it has to do to relieve itself of its bondages and gain enlightenment.

During the period since Lord Mahavira formalised the fundamental principles of Jain philosophy, these have been interpreted in different ways and have given rise to a number of different sects within the Jain faith. To understand the differences that have arisen, we need to look at the history of the faith during the intervening year.

4

Philosophy of Sikhism

Over twenty million Sikhs follow a revealed, distinct, and unique religion born five centuries ago in the Punjab region of northern India. Between 1469 and 1708, ten Gurus preached a simple message of truth, devotion to God, and universal equality. Often mistaken as a combination of Hinduism and Islam, the Sikh religion can be characterized as a completely independent faith:

> *Sikhism rejects idolatry, the caste system, ritualism, and asceticism. It recognizes the equality between both genders and all religions, prohibits the intake of any intoxicants, and encourages an honest, truthful living. Sikhs have their own holy scripture, Guru Granth Sahib. Written, composed, and compiled by the Sikh Gurus themselves, the Guru Granth Sahib serves as the ultimate source of spiritual guidance for Sikhs. While the Sikhs hold their Gurus in high reverence, they are not to be worshipped; Sikhs may only worship God.*

Members of the Sikh community are mainly concentrated in their homeland, Punjab; however, substantial Sikh populations exist throughout the rest of India and the world. Punjabi, a variant of the Hindi language with some Persian influence, is the spoken and written language of the Sikh people. Male members of the Sikh religion use the name, Singh (lion), as their middle or last name, while females use the name, Kaur (princess). Sikhs tend to be industrious and pioneering; this accounts for their general success wherever they live and settle. The hard-working nature of the Sikhs is derived from their religion, which can be best characterized as a faith of unlimited optimism.

Basics of Sikhism

1. Sikhism is the youngest of the World religions. Its history dates back to 1469.
2. Its founder Guru Nanak Dev was born in a village called Talwandi, now known as Nankana Sahib in Pakistan, in 1469.
3. The Sikhs have ten Gurus. It is believed that they all had same soul though they had different bodies, and that it was Guru Nanak Dev's spirit which passed on into his nine successors.
4. The Sikhs call God as 'Waheguru', meaning that God is great. Their common salutation is Sat Siri Akal (God is supreme and is immortal).

"Waheguru Ji Ka Khalsa, Waheguru Ji Ki Fateh"

has a two fold meaning. It denotes a special relationship between God and those who dedicate their lives to His love and service. Also it is the expression of a devotee's faith in the ultimate triumph of Truth over Falsehood. This Sikh salutation means "Khalsa belongs to God and to God alone belongs the Victory".

5. Guru Granth Sahib is the holy book of the Sikhs. It is believed that the tenth Guru, Guru Gobind Singh bestowed upon the Granth the title of the Guru.
6. The Sikhs worship only one Almighty God in his abstract form. They are not allowed to worship any idols, images or photographs.
7. According to the Sikh belief, God is the eternal truth; he is beyond fear, enmity and death. He is unborn and is self illuminated. He himself is the creator, preserver and destroyer.
8. The Sikhs believe that all existence is controlled by one omnipotent, omnipresent and omniscient Lord called by different names: Ishwar, Jehovah, Allah and Waheguru.

Who is a Sikh: A Sikh is any person whose faith consists of belief in One God, the ten Sikh Gurus, the Guru Granth Sahib and other scriptures and teachings of the Sikh Gurus. Additionally, he or she must believe in the necessity and importance of 'Amrit', the Sikh baptism.

God and the Sikhs: According to the Sikh belief, God is all omnipotent, omnipresent and omniscient. The sun, moon/s, wind, fire, water, vegetation and all other things which exist are His witnesses. A Sikh must worship only the abstract form of God. The worship of images or any other object is strictly forbidden.

God is both the creator and the destroyer. He is beyond birth and death. He is both merciful and compassionate. He is beyond fear and enmity. He is self illuminated. He is the Master of all the treasures. All our possessions are a result of His grace. The Sikhs call God as Waheguru, meaning the most wonderful Master.

The belief of the Sikhs in Waheguru is similar to that of Judaism, Christianity and Islam *i.e.*, God is the greatest power, He is supreme, He is the king of kings, He pervades everywhere, He knows the inner thoughts of everyone, He is the giver, He existed before the start of the time, He existed when the time was started, He exists now and He will exist forever.

Relationship with God: The Sikh Gurus called Waheguru as Master and themselves as his servants. In some hymns they called Him as Father, Mother, Friend and Brother as well1. Like Jesus Christ, Guru Gobind Singh, the tenth Guru of the Sikhs, in one of his hymns, called himself as God's son2.

Universality: Sikhism does not believe in asceticism, celibacy or living alone at mountains or in caves or in forests in the search of Truth and God. It also rejects the orders of monasteries. For a Sikh the true life is the life of a householder. Living in a family environment and by serving the community both Truth and God can be realised. Thus it rejects the order of monks (Buddhism and Jainism) and nuns (Christianity).

The Sikh teachings are based on the principles of Fatherhood of God and brotherhood of humankind.

Sikhism rejects the concept of chosen people (as in Judaism) and caste system (as in Hinduism); it also rejects the concept of entering 'Nirvana' without the blessings of God (as in Buddhism and Jainism).

In a Sikh temple people of all the faiths are welcome. The Sikh holy book, Guru Granth Sahib also has in it the hymns composed by both Hindu and Muslim saints of various denominations.

The first five baptised Sikhs, called the beloved ones, were also from both lower and upper Hindu castes. They were the first Khalsa, the pure ones:

Bhai Daya Singh, aged 30, a Khatri from Lahore (Punjab)

Bhai Dharam Singh, aged 33, a Jat from Delhi

Bhai Mohkam Singh, aged 36, a washerman from Dwarka (Gujrat)

Bhai Sahib Singh, aged 37, a barber from Bidar (Karnatak)

Bhai Himmat Singh, aged 39, a water carrier from Puri (Orissa)

Guru Gobind Singh, the tenth prophet of the Sikhs, urged his followers to drop caste symbols after their names and instead write a common surname: Singh, meaning lion, for men and Kaur, meaning princess for women.

The Khalsa

A baptised Sikh is called Khalsa, who must observe and follow strict code of conduct:

1. worship only one Almighty God,
2. recite five prescribed banis (hymns) everyday
3. learn Punjabi language and read Guru Granth Sahib,
4. wear and observe the significance of five Ks: kesh-uncut hair, kanga-a small comb, kara-a stainless steel bracelet, kirpan-a sword and kuchcha-an underwear.
5. live a truthful life and treat all humans as equal

He must not:

1. cut body hair
2. eat kosher meat,
3. smoke, take drugs or intoxicants,
4. have faith in black magic, superstitions, charms and rituals

Rules, Concepts and Commandments: A Sikh lives by the rules made for him by the ten Gurus. The fundamental rules, concepts and the commandments are as follows:

Worship of God

1. Worship only God and no one else.
2. Make worship and prayer a part of your daily life.
3. Do not make images of God, worship him in his abstract form.

Fatherhood of God and Brotherhood of Man

1. Believe that everyone is the child of God.
2. Believe that all human-beings are equal.
3. Do not discriminate on the basis of colour, religion, cast and creed.

Rituals and Superstitions

1. Do not believe in any rituals and superstitions.
2. Do not believe in the worship of images, tombs and graves.

Social and Family Guidelines

1. Do not take alcohol, tobacco and drugs.
2. Do not eat halal meat.
3. Do not eat any food which inflames the passions.
4. Be true to your parents and children.
5. Do not steal.
6. Do not gamble.
7. Love and respect your guests.

The Belief

1. Believe in the oneness of the ten Gurus.
2. Believe in the sovereignty of Guru Granth Sahib
3. Believe in the spiritual authority of the five takhats.

The Nitnem (Daily prayers): The banis which must be recited everyday are:

The bani	*The author*	*Where is it recorded*
Japji Sahib	Guru Nanak Dev	pages 1-8 Guru Granth Sahib
Jap Sahib	Guru Gobind Singh	pages 1-10 Dasam Granth
Swayas	Guru Gobind Singh	pages 13-15Dasam Granth
Rehras	Guru Nanak	pages 8-12 Guru Granth Sahib
	Guru Amardas	
	Guru Amardas	
	Guru Arjan	
Kirtan Sohila	Guru Nanak	pages 12-13 Guru Granth Sahib
	Guru Ramdas	
	Guru Arjan	

The Sikh Prayers and Their Timings: The prayers are classified as: Individual and Collective prayers. The Individual prayers are Morning prayers: Jap ji, Jap Sahib and Sudha Swayas Evening prayer: Rehras Sahib Night time prayer: Kirtan Sohila.

The Congregational prayer is: Morning prayer: Asa di var(composed by Guru Nanak Dev)

In addition to the above prayers a Sikh normally reads the following: At the end of every service or prayer: Anand Sahib (composed by Guru Amardas) At any special occasion or otherwise: Sukhmani Sahib (composed by Guru Arjan) At the time of marriage: Lavan (composed by Guru Ramdas) At the time of cremation: Kirtan Sohila

The main sources of Sikh material to understand the religion in depth are:

1. Guru Granth Sahib, the Sikh holy book
2. Dasm Granth, the holy book of the Tenth Guru Gobind Singh
3. Janam Sakhis, the life stories of the Sikh Gurus

4. Hukamnama, the letter written by the Sikh Gurus
5. Varan Bhai Gurdas, the hymns composed by Bhai Gurdas I & II, contemporaries of Guru Arjan and Guru Gobind Singh Ji respectively.

The Sikh Festivals: The Sikhs celebrate both religious and social festivals. The religious festivals are however called Gurbpurbs

The important Sikh festival calendar is as follows:

Month	***Gurpurb***	***Other festivals***
Jan-Feb	Birthday Guru Harrai	Maghi, Basant
Feb-March	xxxxxxxxxx	Hola Mohalla
March-April	Birthday Guru Angad	Baisakhi
April-May	Birthday Guru Arjan	
	Birthday Guru Tegh Bahadur	
May-June	Birthday Guru Amardas	
	Martyrdom day Guru Arjan	
June-July	Birthday Guru Hargobind	
July-August	Birthday Guru Harkrishen	
August-September	xxxxxxxxxxxxx	Rakhsha Bandan
September-October	Birthday Guru Ramdas	
	Installation Guru Granth Sahib	
October-November	Birthday Guru Nanak	Diwali
November-December	Martyrdom of Guru Tegh Bahadur	
December-January	Birthday Guru Gobind Singh	Lohri

The Sikh Shrine—Gurudwara

1. A Sikh shrine is called a Gurdwara, meaning the doorway to the house of God.
2. The first Gurdwara was built by Guru Nanak Dev at Kartarpur.
3. The Sikh Gurdwaras must have a religious flag, called Nishan Sahib in the front of the Gurdwara.
4. Guru Granth is placed on the far side centre of the hall.
5. There should be no photographs of the Gurus or others in the hall where Guru Granth Sahib is installed.

6. Gurdwaras normally have two halls/rooms. The main hall where Guru Granth Sahib is placed and the second hall where the community kitchen is served.
7. All entrants must take off their shoes, wash their feet and cover their heads before entering the main hall.
8. All Sikh services end with the distribution of parshad (sweet pudding) and langar (dinner/lunch).

Five historical Sikh gurdwaras have been declared as the Sikh Takhats (thrones). These gurdwaras are vested with the power and authority to regulate the religious life of the Sikh nation. The head priests of these shrines constitute a Sikh parliament and they are empowered with executive, legislative and judicial powers regarding the Sikh religious issues. All Sikhs are under the authority of the five takhats. The takhats are as follows:

The Name of the Shrine

The names of the Guru its relates to:

Takhat Akal Takhat: Founded by Guru Hargobind

Takhat Patna Sahib: The birth place of Guru Gobind Singh

Takhat Hazoor Sahib: The place where Guru Gobind Singh breathed his last.

Takhat Kesgarh Sahib: The birth place of the Khalsa

Takhat Damdama Sahib: The place where Guru Gobind Singh composed the second version of Guru Granth Sahib.

All the five takhats relate to the two Gurus who were Saint-soldiers.

The Ten Gurus

Sri Guru Nanak Dev Ji

Sri Guru Angad Dev Ji

Guru Amar Das Ji

Sri Guru Ram Das Sahib Ji

Sri Guru Arjan Dev Ji

Sri Guru Hargobind Sahib Ji

Sri Guru Har Rai Sahib Ji
Sri Guru HarKishan Sahib ji
Sri Guru Tegh Bhadur Sahib Ji
Guru Gobind Singh Ji

Skhism and the Sikh Gurus

Guru Nanak Dev Ji (1469-1539): The Sikhs had ten prophets called the Gurus. The time-period of the history of the Gurus ranged from 1469 A.D, when Guru Nanak, the founder of the Sikhism was born, to 1708 A.D, when the last prophet, Guru Gobind Singh left this mortal world for his heavenly abode (239 years).*Guru Nanak,* the founder of the Sikh religion was born in a Hindu family of Kshatriya caste. He revolted against that order when he was only 13 years old. God spoke to him when he was 38. He was taken to God's abode where God gave him two Commandments known as 'Moolmantar' and 'Sloak'.

Moolmantar: *There is only one God He is the Truth (permanent being) He is the Creator, He is devoid of fear He is devoid of enmity He is beyond death He is not born He is self-illuminated He is the master of all the gifts.*

God existed before the start of the time God existed when the time started. God exists now and, God will always exist. Guru Nanak was appointed as God's latest prophet and was asked to spread the name of God and the rules of both divinity and morality.Guru Nanak travelled as far as Tibet in the north, Sri Lanka in the south, Mecca in the west and Dhaka in the east to deliver God's message. He was welcomed wherever he went. He met both kings and robbers, gave them the message of God and reformed them. He is the only prophet in the world who did not meet any violent opposition and was not harmed by the enemies.

In fact he had no enemies. (compare him with: Moses and the Pharaos; Jesus and the Jewish clergy and the Romans; Ram and Ravan; Krishen and Kans; Mohammed and the Meccan pagans).Guru Nanak was revered by both the Hindus and the Muslims. He was received with respect and folded hands by

Babar, the Muslim Emperor of India and Shivnabh, the Hindu King of Sri Lanka; he was honoured by the clergy of both the Hindus and the Muslims. Guru Nanak was a family man, was married and had two sons. While living with his father he looked after the family fields and the cattle; staying with his sister, Babe Nanki, he worked as a store keeper with the local ruler; and for the last 17 years of his life, he tilled his own fields at Kartarpur, a town founded by him.

There are 41 Sikh shrines founded in his memory.For 14 years, from 1507-1510 he travelled around the then known world and spread the message of God. In history, he is the most travelled of all the known prophets. (Jesus travelled in the central plains of Palestine; Mohammed travlled from Mecca to Medina and back to Mecca; Moses travelled from Egypt to outskirts of Palestine; Ram travelled from Ayodhaya to Sri Lanka; Krishna travelled from Mathura to Dwarka).Guru Nanak died at the age of 70. His 974 hymns are recorded in the Sikh holy book.

The Prophets

2nd Guru-10th Guru

Guru Angad (1504-1552): was the second Guru of the Sikhs. He was a disciple of Guru Nanak and was chosen as his successor after being put to a great many tests. He became Guru at the age of 35 and his pontificate lasted for 13 years. He rationalised the Panjabi language and gave to it a new grammar. He also encouraged his followers to look after their health. He gave them instructions to have a balanced diet and regular exercises. He built many wrestling arenas and encouraged his followers to participate in wrestling competitions Like Guru Nanak, he founded a new town and named it Khadur. There are 2 Gudwaras built in his memory, and there are in Guru Granth Sahib, 65 hymns composed by him. He died at the age of 48.

Guru Amardas (1479-1574): Became the third Guru at the age of 73. For twelve years he personally served Guru Angad. He walked daily for 5 miles to fetch water for the Guru's bath. He was chosen from amongst many by Guru Angad as his successor. His pontificate lasted for 17 years. He inculcated amongst his followers, the spirit of Service to humankind and to God Like both Guru Nanak and Guru Angad he was a very simple man. There are 4 Gurdwaras related to his memory and there are, in Guru Granth-Sahib, 907 hymns composed by him. He died at the age of 90.

Guru Ramdas (1534-1581): Became Guru when he was 40. His pontificate lasted for only 7 years. He was a son-in-law of Guru Amardas. He re-organised the Sikh Church and founded the city of Amritsar.He died at the age of 47. There are 3 Gurdwaras built in his memory and there are 679 of his hymns recorded in Guru Granth Sahib.

Guru Arjan (1563-1606): became Guru at the young age of 18. He was the youngest son of Guru Ramdas, He built the Golden Temple and compiled the Sikh holy book. He also founded the city of Taran Taran. He is the first martyr of the Sikh history. He died at the age of 43, There are 12 Gurdwaras built in his memory and there are 2,218 of his hymns recorded in Guru Granth Sahib.

Guru Hargobind (1595-1644): Became Guru at the age of 11. He was the only son of Guru Arjan. His pontificate was longest amongst all the Gurus, it lasted for 38 years. He was the first Guru to fight with the Mughals against their injustice and tyranny.. He wore two swords, symbolising Miri and Piri, royalty and saintliness. He organised early morning Sikh choirs. He founded the city of Kiratpur3. There are 16 Gudwaras built in his memory. He travelled upto Kashmir in the north to spread the message of the house of Nanak. He died at the age of 49. He did not write any hymns.

Guru Harrai (1630-1661): was a grandson of Guru Hargobind. He became Guru at the age of 14 and remained Guru for 17 years. He built many clinics for both sick human beings and animals/birds. He was expert in Aryuvedic medicine. Most of the time he lived in Nahan and preached the divine message over there. There are 3 Gurdwaras dedicated to his memory. He died at the age of. He did not write any hymns.

Guru Harkrishen (1656-1664): was the youngest son of Guru Harrai. He became Guru at the tender age of 5 and died at the age of 8. Thus his pontificate lasted for only 3 years. He took over the sufferings of the people of Delhi over himself and saved them from effects of smallpox epidemic (compare this sacrifice with the Christian saying, "Jesus died for our sins"). He instructed his followers to build schools for religious education. There are 4 Gurdwaras related to him. He did not write any hymns. Most of the modern Sikh Schools are named after him.

Guru Tegh Bahadur (1621-1675): was a grand uncle of Guru Harkrishan. He became Guru at the age of 43 and remained Guru until the age of 54. He travelled towards east of India up to Dhaka to spread the message of the house of Nanak. He offered himself for martyrdom for saving the Hindu religion. He gave his life but saved the annihilation of Hindu religion by the then Mughal Emperor, Aurangzeb. He was beheaded in Delhi at a place known as 'Chandni Chowk' He founded the city of Anandpur. There are 16 Gurdwaras built in his memory and there are 115 hymns recorded under his name in Guru Granth Sahib.

Guru Gobind Singh (1666-1708): became Guru at the age of 9. He was the only son of Guru Tegh Bahadur. Like his grand-father Guru Hargobind, he also had to resort to sword to protect the young Sikh nation from an onslaught of the Mughals. He initiated a new baptism and called it 'Amrit' He created the order of Khalsa (Saint-soldiers) and prescribed the compulsion of wearing the 5 Ks. He wrote hymns which were later collected by one of his followers, Bhai Mani Singh, at the orders of his widow Mata Sundri. He declared the mission of Guru Nanak completed. He passed on the spiritual authority of the Sikhs to the Sikh holy book, and called it Guru Granth Sahib. He passed on the temporal authority of the Sikh to the Khalsa. He died at the age of 42. His hymns are preserved in the Granth called Dasam Granth. He was the last prophet (Guru) of the Sikhs.

The Theory of one Sprit: The Sikhs believe that all the ten Gurus had the same spirit. This is one of the fundamental beliefs of the Sikhs. Guru Gobind Singh had recorded in one of his hymns that though after Nanak came Angad and then Amardas and then Ramdas, Arjan, Hargobind, Harrai, Harkrishen and Tegh Bahadur, but they all had the same spirit. They looked different for they had different bodies but their spirit, the inner self, was the same. (Compare it with Christian belief of Holy Spirit).

1. Guru Granth Sahib-Guru Arjan Dev: "Tu mera pita tu hai mera mata, tu mera bandhu tu mera bharata".
2. Dasam Granth-Bachitar Natak: "Tu nae apna sut mujae niwaja".
3. *It is at this place that the Sikhs immerse the ashes of their dead. The Gurdwara is called 'Patal Puri'.*

Bibliography

Bhardwaj, Surinder Mohan : *Hindu Places of Pilgrimage in India : A Study in Cultural Geography,* University of California Press, 1973.

Bhattacharya, Narendra Nath : *A Dictionary of Hindu Mythology,* Munshiram Manoharlal, Delhi, 2001.

Biardeau, Madeline : *Hinduism : The Anthropology of a Civilization,* Oxford UP, 1997.

Brockington, J. L. : *Righteous Rama : The Evolution of an Epic,* Oxford, 1984.

——————: *The Sacred Thread : A Short History of Hinduism,* Oxford UP, 1992.

Bryant, Kenneth E. : *Poems to the Child-God : Structures and Strategies in the Poetry of Surdas,* University of California Press, 1979.

Buhler, Georg : *The Laws of Manu,* Clarendon Press, Oxford, 1886.

Chaudhuri, Nirad C. : *Hinduism : A Religion to Live By,* Oxford UP, 1979.

Danielou, Alain : *The Complete Kama Sutra,* Park Street Press, Rochester, Vermont, 1994.

Dasgupta, S. N. : *Hindu Mysticism,* Frederick Ungar, New York, 1959.

——————: *History of Indian Philosophy,* Cambridge UP, 1922-55.

Dimock, Edward and Denise Levertov : *In Praise of Krishna : Songs from the Bengali,* Anchor Books, New York, 1967.

Dowson, John : *A Classical Dictionary of Hindu Mythology and Religion, Geography, History, and Literature*, Routledge & Kegan Paul, 1957.

Edgerton, Franklin : *The Bhagavad Gita,* Harvard UP and Oxford UP, 1944.

Eliade, Mircea : *Yoga : Immortality and Freedom*, Princeton/ Bollingen, 1970.

Farquhar, J. N. : *An Outline of the Religious Literature of India,* Oxford UP, 1920.

Fuller, C. J. : *The Camphor Flame : Popular Hinduism and Society in India,* Princeton UP, 1992.

Gandhi, M. K. : *Hindu Dharma*, Navajivan, Ahmedabad, 1950.

———— : *The Bhagavad Gita According to Gandhi*, Berkeley Hills Books, 2000.

———— : *What is Hinduism?* National Book Trust, Delhi, 1994.

Goldman, Robert : *The Ramayana of Valmiki,* Princeton UP, 1984.

Griffith, Ralph T. H. : *The Hymns of the Rigveda,* E.J. Lazarus and Co., Benares, 1920.

Growse, F. S. : *The Ramayana of Tulasidasa,* Motilal Banarsidass, Delhi, 1995.

———— : *The Ramayana of Tulasidasa,* Motilal Banarsidass, Delhi, 1995.

Gupta, Mahendranath : *The Gospel of Sri Ramakrishna*, Ramakrishna-Vivekananda Center, New York, 1969.

Hiriyanna, M. : *The Essentials of Indian Philosophy,* George Allen & Unwin, London, 1948.

Hume, Robert Ernest : *The Thirteen Principal Upanishads,* Oxford UP, 1983.

Kane, P.V. : *History of Dharmasastra,* Bhandarkar Oriental Research Institute, Poona, 1930.

Kapoor, Jagdish Chander : *Bhagavad-Gita, an International Bibliography of 1785-1979 Imprints*, Garland Pub., New York, 1983.

Keith, A. B. : *The Religion and Philosophy of the Vedas and Upanishads,* Motilal Banarsidass, Delhi, 1970.

Klostermaier, Klaus K. : *A Concise Encyclopaedia of Hinduism,* Oneworld, Oxford, 1998.

Leslie, I. J. : *Roles and Rituals for Hindu Women,* Pinter Publishers, London, 1991.

Lipner, Julius : *Hindus : Their Religious Beliefs and Practices,* Routledge, 1994.

Lopez, Donald S. : *Religions of India in Practice,* Princeton UP, 1995.

Mahadevan, T. M. P. : *Outlines of Hinduism,* Chetna, Bombay, 1971.

Mani, Vettam : *Puranic Encyclopaedia,* Motilal Banarsidass, Delhi, 1984.

Michell, George : *The Hindu Temple: An Introduction to its Meaning and Form,* University of Chicago Press, 1988.

Miller, Barbara Stoller: *The Bhagavad-Gita,* Bantam Books, 1986.

Mirabai : *The Devotional Poems of Mirabai,* Motilal Banarsidass, Delhi, 1980.

Narayan, R. K. : *The Mahabharata,* Viking, New York, 1978.

Nikhilananda, Swami : *The Upanishads,* George Allen & Unwin, London, 1963.

Pande, G. C. : *Foundations of Indian Culture,* Motilal Banarsidass, Delhi, 1983.

Panikkar, Raimundo : *The Vedic Experience : An Anthology of the Vedas for Modern Man and Contemporary Celebration,* Motilal Banarsidass, New Delhi, 1994.

Prasad, R. C. : *Tulasidasa's Shriramacharitamanasa-The Holy Lake of the Acts of Rama,* Motilal Banarsidass, Delhi, 1999.

Radhakrishnan, S. : *The Hindu View of Life,* MacMillan, New York, 1927.

——————— : *The Bhagavad Gita,* George Allen & Unwin, London, 1948.

Radhakrishnan, S. and Charles A. Moore : *A Sourcebook in Indian Philosophy,* Princeton, 1957.

Ramanujan, A. K. : *Hymns for Visnu by Nammalvar,* Princeton, 1981.

Rocher, Ludo : *The Puranas*, Otto Harrassowitz, Wiesbaden, 1986.

Schelling, Andrew : *For Love of the Dark One : Songs of Mirabai,* Hohm Press, 1998.

Sen, K. M. : *Hinduism,* Penguin, 1961.

Shastri, J. L. : *Ancient Indian Tradition and Mythology,* Motilal Banarsidass, Delhi, 1970.

Shukla, D. N. : *Vastu-Shastra,* Munshiram Manoharlal, Delhi, 1993.

Singer, Milton : *Krishna : Myths, Rites, and Attitudes,* University of Chicago Press, 1968.

Urban, Hugh B. : *Songs of Ecstasy: Tantric and Devotional Songs from Colonial Bengal,* Oxford, 2002.

Walker, Benjamin : *The Hindu World*, Frederick A. Praeger, New York:, 1968.

Williams, R. B. : *A New Face of Hinduism : The Swaminarayan Religion,* Cambridge UP, 1984.

Williams, Raymond Brady : *A Sacred Thread : Modern Transmission of Hindu Traditions in India and Abroad,* Columbia UP, 1992.

Woods, James Houghton : *The Yoga-System of Patanjali,* Motilal Banarsidass, Delhi, 1966.

Zaehner, R. C. : *Hinduism,* Oxford UP, 1962.

Zimmer, Heinrich : *Philosophies of India*, Joseph Campbell, Princeton/Bollingen, 1951.

Index

S

T

❑❑❑